Annals of Mathematics Studies

Number 166

Lectures on Resolution of Singularities

János Kollár

PRINCETON UNIVERSITY PRESS
PRINCETON AND OXFORD
2007

Published by Princeton University Press
41 William Street, Princeton, New Jersey 08540

In the United Kingdom: Princeton University Press
3 Market Place, Woodstock, Oxfordshire OX20 1SY

Library of Congress Cataloging-in-Publication Data

Kollár, János.
    Lectures on resolution of singularities / János Kollár.
        p. cm. — (Annals of mathematics studies ; 166)
    Includes bibliographical references and index.
    ISBN-13: 978-0-691-12922-8 (acid-free paper)
    ISBN-10: 0-691-12922-3 (acid-free paper)
    ISBN-13: 978-0-691-12923-5 (pbk. : acid-free paper)
    ISBN-10: 0-691-12923-1 (pbk. : acid-free paper)
    1. Singularities (Mathematics). I. Title.

QA614.58.K68 2007
516.3'5–dc22   2006050554

British Library Cataloging-in-Publication Data is available

This book has been composed in Computer Modern in LaTeX

The publisher would like to acknowledge the authors of this volume for
providing the camera-ready copy from which this book was printed.

Printed on acid-free paper. ∞

press.princeton.edu

Printed in the United States of America

10 9 8 7 6 5 4 3 2 1

Contents

Introduction

A complex algebraic variety X is a subset of complex affine n-space $\mathbb{C}^n$ or of complex projective n-space $\mathbb{CP}^n$ defined by polynomial equations.

A point $x \in X$ is called a *smooth* point if, up to a complex analytic local coordinate change, X looks like a linear subspace near x. Otherwise, a point is called *singular*. At singular points X may have self-intersections, it may look like the vertex of a cone, or it can be much more complicated.

Roughly speaking, resolution of singularities asserts that an arbitrary singular variety X can be parametrized by a smooth variety X'. That is, all the points of X' are smooth, and there is a surjective and proper map $f : X' \to X$ defined by polynomials.

The whole history of algebraic geometry is intertwined with the development of resolution of singularities. One of the first deep results in algebraic geometry is Newton's proof of resolution for curves in the complex plane $\mathbb{C}^2$, that is, for zero sets of two variable polynomials $f(x, y) = 0$.

The next two centuries produced many more proofs of resolution for curves, and by the early part of the twentieth century the resolution of algebraic surfaces was also settled. See [Bli23, Zar71, Gar88] for some historical accounts.

After much effort, Zariski proved resolution for 3-folds [Zar44], and finally Hironaka settled the general case in a 218-page paper in 1964 [Hir64].

Since then resolution has had an unusual role in algebraic geometry. On the one hand, it is a basic result and almost everyone working in the field uses it; many of us use it regularly. On the other hand, the proof was generally viewed as too long and complicated to fathom in detail, and few people actually read all of it. (I must confess that I have not been among these few until recently.)

The lingering perception that the proof of resolution is very hard gradually diverged from reality. While Hironaka's original arguments are indeed very subtle and lengthy, during the last forty years a small group of initiates has been improving and simplifying the proof. In particular, all the technical machinery has been removed.

I gave a graduate course devoted to resolution of singularities at Princeton University during the 2004/05 academic year. The first semester gave ample time to explore many different approaches for curves and surfaces. It is quite interesting to see how the nice methods for curves become much harder for surfaces before a clear and simple proof emerges for the general case. Finally it is feasible to prove resolution in the last two weeks of a beginning algebraic geometry course.

Chapter 1 is devoted to resolution of curve singularities. It was very enjoyable to work through thirteen methods, read some old papers, and see the roots of many later techniques. This chapter is very elementary, and many of the proofs could be presented in a first course on algebraic geometry.

Chapter 2 studies resolution for surfaces. Several pretty ideas work for surfaces but seem to fall short in higher dimensions. In addition, the proofs need more technical background.

By contrast, the methods of the general case presented in Chapter 3 are again elementary. The actual proof occupies only thirty pages in Sections 7–13, and most of Chapter 3 is devoted to motivation and examples. The approach is based on the recent work of Włodarczyk [Wło05], which in turn is built on the earlier results of Hironaka [Hir64, Hir77], Giraud [Gir74] and Villamayor [Vil89, Vil92]. See the introduction to Chapter 3 for an overview and a description of the new features of the proof.

Other important advances in the Hironaka method are in [BM97, BV01, EH02], whose connections with the present proof are more indirect.

It is worth noting that before the appearance of [Hir64] it was not obvious that resolutions existed, much less that they could be achieved by repeatedly blowing up smooth subvarieties. The 1950s saw several wrong proofs, and even Zariski—the grandfather of modern resolution—seems to have doubted at some point that the Hironaka method would ever work. See [Rei00] for the history of these years.

This short book is aimed at readers who are interested in the subject and would like to understand *one* proof of resolution in characteristic 0 but do not (yet) plan to get acquainted with every aspect of the theory. Any thorough treatment of resolutions should encompass many other topics that are barely mentioned here. Some of these are the following:

- Abhyankar's methods in positive characteristic [Abh66],
- Bierstone and Milman's work developing Hironaka's "idealistic" paper [Hir77, BM89, BM91, BM97, BM03],
- Bravo and Villamayor's strengthening of resolution [BV01],
- computer implementations [BS00b, BS00a, FKP05],
- de Jong's construction of alterations and their application to resolutions [dJ96, BP96, AdJ97, AW97, Par99],
- Deligne's theory of simplicial resolutions [Del71, GNAPGP88],
- desingularization of vector fields [Sei68, Can87, Can04],

- Encinas and Hauser's method using mobiles [EH02, Hau03],
- Lipman's work on excellent surfaces [Lip78, Art86a],
- Mumford's semi-stable reduction theorem and its higher dimensional versions [KKMSD73, Kar00, AK00, Cut02],
- Néron desingularization [Nér64, Art86b, BLR90],
- simultaneous resolution [Tei82, Lau83, KSB88, Lip00, Vil00, GP03, ENV03],
- toric and toroidal methods [KKMSD73, AMRT75], and
- Włodarczyk's results on factorization of birational maps and its variants [Wło00, Wło03, HK00, AKMW02, Kar05].

ACKNOWLEDGMENTS. These notes owe a lot to the active participation of my audience at Princeton University. The comments of S. Grushevsky, A. Hogadi, D. Kim, F. Orgogozo and C. Xu have been especially helpful. Various parts of the manuscript have been in circulation, and I received many useful corrections and suggestions from D. Abramovich, J. Lipman, K. Matsuki, Ch. Rotthaus and E. Szabó. I am especially grateful for the long lists of comments from E. Bierstone, H. Hauser, S. Kleiman, P. Milman and J. Włodarczyk.

The final version of Chapter 3 was completed in connection with a lecture series at the University of Utah in 2006. I thank A. Bertram, C. Hacon and J. McKernan for good questions and comments, N. Beebe for TeX help and the Department of Mathematics for its hospitality.

Partial financial support was provided by the National Science Foundation under grant numbers DMS-0200883 and DMS-0500198.

CHAPTER 1

Resolution for Curves

Resolution of curve singularities is one of the oldest and prettiest topics of algebraic geometry. In all likelihood, it is also completely explored.

In this chapter I have tried to collect all the different ways of resolving singularities of curves. Each of the thirteen sections contains a method, and some of them contain more than one. These come in different forms: solving algebraic equations by power series, normalizing complex manifolds, projecting space curves, blowing up curves contained in smooth surfaces, birationally transforming plane curves, describing field extensions of Laurent series fields and blowing up or normalizing 1-dimensional rings.

By the end of the chapter we see that the methods are all interrelated, and there is only one method to resolve curve singularities. I found, however, that these approaches all present a different viewpoint or technical twist that is worth exploring.

1.1. Newton's method of rotating rulers

Let $F(x, y)$ be a complex polynomial in two variables. We are interested in finding solutions of $F = 0$ in the form $y = \phi(x)$, where ϕ is some type of function that we are right now unsure about.

Following the classical path of solving algebraic equations, one might start with the case where $\phi(x)$ is a composition of polynomials, rational functions and various mth roots of these. As in the classical case, this will not work if the degree of F is 5 or more in y.

One can also try to look for power series solutions, but simple examples show that we have to work with power series with fractional exponents. The equation $y^m = x + x^2$ has no power series solutions for $m \geq 2$, but it has fractional power series solutions for any $\epsilon^m = 1$ given by

$$y = \epsilon x^{1/m} \left(1 + \sum_{j \geq 1} \binom{1/m}{j} x^j \right) \quad \text{for } i = 1, \ldots, m.$$

As a more interesting example, $y^m - y^n + x = 0$ for $m > n$ also has a fractional power series solution

$$y = \sum_{i \geq 1} a_i x^{i/n},$$

where $a_1 = 1$ and the other a_i are defined recursively by

$$n \cdot a_s = \text{coefficient of } x^{(s+n-1)/n} \text{ in } \left(\sum_{i=1}^{s-1} a_i x^{i/n} \right)^m - \left(\sum_{i=1}^{s-1} a_i x^{i/n} \right)^n.$$

After many more examples, we are led to look for solutions of the form

$$y = \sum_{i=0}^{\infty} c_i x^{i/M},$$

where M is a natural number whose dependence on $\deg F$ we leave open for now. These series, though introduced by Newton, are called *Puiseux series*. We encounter them later several times.

THEOREM 1.1 (Newton, 1676). *Let $F(x,y)$ be a complex polynomial or power series in two variables. Assume that $F(0,0) = 0$ and that y^n appears in $F(x,y)$ with a nonzero coefficient for some n. Then $F(x,y) = 0$ has a Puiseux series solution of the form*

$$y = \sum_{i=1}^{\infty} c_i x^{i/N}$$

for some integer N.

REMARK 1.2. (1) The original proof is in a letter of Newton to Oldenburg dated October 24, 1676. Two accessible sources are [New60, pp.126–127] and [BK81, pp.372–375].

(2) Our construction gives only a formal Puiseux series; that is, we do not prove that it converges for $|x|$ sufficiently small. Nonetheless, if F is a polynomial or a power series that converges in some neighborhood of the origin, then any Puiseux series solution converges in some (possibly smaller) neighborhood of the origin. This is easiest to establish using the method of Riemann, to be discussed in Section 1.2.

(3) By looking at the proof we see that we get n different solutions (when counted with multiplicity).

The proof of Newton starts with a graphical representation of the "lowest order" monomials occurring in F. This is now called the Newton polygon.

DEFINITION 1.3 (Newton polygon). Let $F = \sum a_{ij} x^i y^j$ be a polynomial or power series in two variables. The *Newton polygon* of f (in the chosen coordinates x and y) is obtained as follows.

In a coordinate plane, we mark the point (i,j) with a big dot if $a_{ij} \neq 0$. Any other monomial $x^{i'} y^{j'}$ with $i' \geq i, j' \geq j$ will not be of "lowest order" in any sense, so we also mark these. (In the figures these markings are invisible, since I do not want to spend time marking infinitely many uninteresting points.)

The Newton polygon is the boundary of the convex hull of the resulting infinite set of marked points.

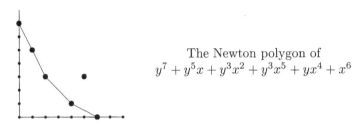

The Newton polygon of
$$y^7 + y^5x + y^3x^2 + y^3x^5 + yx^4 + x^6$$

Assume now that F contains a nonzero term $a_{0n}y^n$ and n is the smallest possible. This means that the Newton polygon has a corner on the y-axis at the point $(0, n)$. Look at the nonvertical edge of the Newton polygon starting at $(0, n)$. Let us call this the *leading edge* of the Newton polygon. (As Newton explains it, we put a vertical ruler through $(0, n)$ and rotate it till it hits another marked point—hence, the name of the method.)

1.4 (Proof of (1.1)). We construct the Newton polygon of F and concentrate on its leading edge.

If the leading edge is horizontal, then there are no marked points below the $j = n$ line, and hence, y^n divides F and $y = 0$ is a solution.

Otherwise, the extension of the leading edge hits the x-axis at a point which we write as nu/v where u, v are relatively prime. The leading edge is a segment on the line $(v/u)i + j = n$. In the diagram below the leading edge hits the x-axis at $7/2$, so $u = 1$ and $v = 2$.

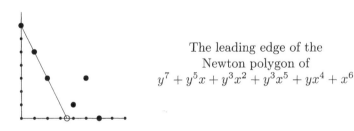

The leading edge of the
Newton polygon of
$$y^7 + y^5x + y^3x^2 + y^3x^5 + yx^4 + x^6$$

We use induction on the leading edge, more precisely, on its starting point $(0, n)$ and on its steepness v/u.

Our aim is to make a coordinate change and to obtain another polynomial or power series $F_1(x_1, y_1)$ with leading edge starting at $(0, n_1)$ and steepness v_1/u_1 such that

- either $n_1 < n$,

- or $n_1 = n$, and $v_1/u_1 < v/u$.

Moreover, we can write down a Puiseux series solution of $F(x, y) = 0$ from a Puiseux series solution of $F_1(x_1, y_1)$.

Then we repeat the procedure. The first case can occur at most n-times, so eventually the second case happens all the time. We then construct a Puiseux series solution from this infinite sequence of coordinate transformations.

In order to distinguish the two cases, we consider the terms in F that lie on the leading edge

$$f(x, y) := \sum_{(v/u)i+j=n} a_{ij} x^i y^j,$$

and we think of this as the "lowest terms" of F. If $a_{ij} x^i y^j$ is a nonzero term in $F(x, y)$, then $(v/u)i + j \geq n$, so $f(x, y)$ indeed consists of the lowest degree terms in F if we declare that $\deg x = v/u$ and $\deg y = 1$.

In the above example, $f(x, y) = y^7 + y^5 x + y^3 x^2$.

Note that $(v/u)i + j = n$ has an integer solution only if $v | n - j$; thus we obtain the following.

Claim 1.4.1. We can write

$$f(1, y) = \sum_{0 \leq k \leq n/v} a_k y^{n-kv}.$$

In particular, if $v \neq 1$, then $f(1, y)$ does not contain the term y^{n-1} and so $f(1, y)$ is not an nth power of a linear form.   □

We distinguish the two cases based on how $f(1, y)$ factors.

Case 1. $f(1, y)$ is not an nth power.

Let α be a root of $f(1, y)$ with multiplicity $n_1 < n$. Then we make the substitutions

$$x := x_1^v, \quad y = y_1 x_1^u + \alpha x_1^u.$$

Note that if $a_{ij} x^i y^j$ is a nonzero term in $F(x, y)$, then $(v/u)i + j \geq n$; thus

$$a_{ij} x^i y^j = a_{ij} x_1^{vi+uj} (y_1 + \alpha)^j$$

and $vi + uj \geq nu$ with equality only if $(v/u)i + j = n$. Thus $F(x_1^v, y_1 x_1^u + \alpha x_1^u)$ is divisible by x_1^{nu}, and we set

$$F_1(x_1, y_1) := x_1^{-nu} F(x_1^v, y_1 x_1^u + \alpha x_1^u).$$

Note furthermore that

$$F_1(0, y_1) = f(1, y_1 + \alpha),$$

and so $y_1^{n_1}$ appears in F_1 with nonzero coefficient.

Furthermore, any Puiseux series solution $y_1 = \phi(x_1)$ of $F_1 = 0$ gives a Puiseux series solution

$$y = \phi(x^{1/v}) x^{u/v} + \alpha x^{u/v}$$

of $F = 0$.

Case 2. $f(1, y)$ is an nth power.

By (1.4.1) this can happen only for $v = 1$. Write $f(1, y) = c(y - \alpha)^n$, and make a coordinate change

$$x = x_1, \; y = y_1 + \alpha x_1^u.$$

Under this transformation, $x^i y^j$ becomes a sum of monomials $x_1^{i'} y_1^{j'}$, where $(1/u)i' + j' = (1/u)i + j$. Thus we do not get any new terms below the leading edge of f, and we kill every monomial on the leading edge save y^n, which is now y_1^n.

Hence $n_1 = n$, but the leading edge of the Newton polygon of

$$F_1(x_1, y_1) := F(x_1, y_1 + \alpha x_1^u)$$

is less steep than the leading edge of the Newton polygon of $F(x, y)$.

Next we repeat the procedure with $F_1(x_1, y_1)$ to get $F_2(x_2, y_2)$ and so on.

As we noted, the only remaining question is, what happens when the second case happens infinitely often. This means that we have an infinite sequence of coordinate changes

$$y_s = y_{s+1} + \alpha_{s+1} x_s^{u_{s+1}}, \quad y_{s+1} = y_{s+2} + \alpha_{s+2} x_s^{u_{s+2}}, \dots \;.$$

Here $u_{s+1} < u_{s+2} < \cdots$; thus we can view this sequence as converging to a single power series substitution

$$y_s = y_\infty + \alpha_{s+1} x_s^{u_{s+1}} + \alpha_{s+2} x_s^{u_{s+2}} + \cdots,$$

and then

$$F_s(x_s, y_\infty + \alpha_{s+1} x_s^{u_{s+1}} + \alpha_{s+2} x_s^{u_{s+2}} + \cdots) = y_\infty^n \text{(invertible power series)},$$

giving the power series solution $y_s = -(\alpha_{s+1} x_s^{u_{s+1}} + \alpha_{s+2} x_s^{u_{s+2}} + \cdots)$. □

1.2. The Riemann surface of an algebraic function

The resolution of singularities of analytic curves is due to Riemann. When he constructs the Riemann surface of a function, he goes directly to the smooth Riemann surface, bypassing the singular model; see [Rie90, pp.39–41]. His method is essentially the one given below.

In more contemporary terminology, here is the result.

THEOREM 1.5 (Riemann, 1851). *Let $F(x, y)$ be an irreducible complex polynomial and $C := (F(x, y) = 0) \subset \mathbb{C}^2$ the corresponding complex curve. Then there is a 1-dimensional complex manifold $\bar{C}$ and a proper holomorphic map*

$$\sigma : \bar{C} \to C,$$

which is a biholomorphism except at finitely many points.

Proof. Since F is irreducible, F and $\partial F/\partial y$ have only finitely many points $\Sigma \subset C$ in common. By the implicit function theorem, the first coordinate projection $\pi : C \to \mathbb{C}$ is a local analytic biholomorphism on $C \setminus \Sigma$.

We start by constructing a resolution for a small neighborhood of a point $p \in \Sigma$. For notational convenience assume that $p = 0$, the origin.

Let $B_\epsilon \subset \mathbb{C}^2$ denote the ball of radius ϵ around the origin. By choosing ϵ small enough, we may assume that $C \cap (y = 0) \cap B_\epsilon = \{0\}$.

Next, by by choosing η small enough, we can assume that

$$\pi : C \cap B_\epsilon \cap \pi^{-1}(\Delta_\eta) \to \Delta_\eta$$

is proper and a local analytic biholomorphism except at the origin, where $\Delta_\eta \subset \mathbb{C}$ is the disc of radius η. Set

$$C_\eta := C \cap B_\epsilon \cap \pi^{-1}(\Delta_\eta) \quad \text{and} \quad C_\eta^* := C_\eta \setminus \{0\}.$$

We thus conclude that

$$\pi : C_\eta^* \to \Delta_\eta^* \quad \text{is a covering map.}$$

The fundamental group of Δ_η^* is $\mathbb{Z}$; thus for every m, the punctured disc Δ_η^* has a unique connected covering of degree m, namely,

$$\rho_m : \Delta_1^* \to \Delta_\eta^* \quad \text{given by } z \mapsto \eta z^m.$$

Let $C_{\eta,i}^* \subset C_\eta^*$ be any connected component and m_i the degree of the covering $\pi : C_{\eta,i}^* \to \Delta_\eta^*$. We thus have an isomorphism of coverings

$$
\begin{array}{ccc}
\Delta_1^* & \xrightarrow{\sigma_i^*} & C_{\eta,i}^* \\
{\scriptstyle \rho_{m_i}} \searrow & & \swarrow {\scriptstyle \pi} \\
& \Delta_\eta^* &
\end{array}
$$

More precisely, topology tells us only that σ_i^* is a homeomorphism. However, the maps ρ_{m_i} and π are local analytic biholomorphisms; thus we can assert that σ_i^* is a homeomorphism that is also a local analytic biholomorphism, and hence a global analytic biholomorphism.

The image of σ_i^* lands in the ball B_ϵ, and hence the coordinate functions of σ_i^* are analytic and bounded on Δ_1^*. Thus by the Riemann extension theorem, σ_i^* extends to a proper analytic map

$$\sigma_i : \Delta_1 \to C_\eta.$$

Doing this for every connected component $C_{\eta,i}^* : i \in I$, we obtain a proper analytic map

$$\sigma : \coprod_{i \in I} \Delta_1 \to C_\eta \quad \text{such that} \quad \sigma^* : \coprod_{i \in I} \Delta_1^* \to C_\eta^*$$

is an isomorphism, where $\coprod_{i \in I}$ denotes disjoint union.

This proves the local resolution for complex algebraic plane curves.

To move to the global case, observe that $\Sigma \subset C$ is a discrete subset. Thus for each $p_i \in \Sigma$ we can choose disjoint open neighborhoods $p_i \in C_i \subset C$. By further shrinking C_i, we have proper analytic maps $\sigma_i : \bar{C}_i \to C_i$, where $\bar{C}_i$ is a disjoint union of open discs and σ_i is invertible outside the singular point $p_i \in C_i$.

We can thus patch together the "big" chart $C \setminus \Sigma$ with the local resolutions $\bar{C}_i$ to get a global resolution $\bar{C}$. $\square$

1.6 (Puiseux expansion). The resolution of the local branches

$$\sigma_i : \Delta_1 \to C_{\eta,i}$$

is given by a power series on Δ_1, and the local coordinate on Δ_1 can be interpreted as x^{1/m_i}.

Thus we obtain that each local branch $C_{\eta,i}$ has a parametrization by a convergent Puiseux series

$$y = \sum_{j=0}^{\infty} a_j x^{j/M}, \quad \text{where } M = m_i \leq \deg_y F.$$

REMARK 1.7. There is lot more to Puiseux expansions than the above existence theorems.

Let $0 \in C \subset \mathbb{C}^2$ be a curve singularity and S_ϵ^3 a 3-sphere of radius ϵ around the origin. Then $C \cap S_\epsilon^3$ is a real 1-dimensional manifold for $0 < \epsilon \ll 1$, and up to diffeomorphism, the pair $(C \cap S_\epsilon^3 \subset S_\epsilon^3)$ does not depend on ϵ. It is called the *link* of $0 \in C$.

$C \cap S_\epsilon^3$ is connected iff C is analytically irreducible, in which case it is called a *knot*. One can read off the topological type of this knot from the vanishing of certain coefficients of the Puiseux expansions. See [BK81, Sec.8.4] for a lovely treatment of this classical topic.

The resolution problem for 1-dimensional complex spaces can be handled very similarly. The final result is the following.

THEOREM 1.8. *Let C be a 1-dimensional reduced complex space with singular set $\Sigma \subset C$. Then there is a 1-dimensional complex manifold $\bar{C}$ and a proper holomorphic map*

$$\sigma : \bar{C} \to C$$

such that $\sigma : \sigma^{-1}(C \setminus \Sigma) \to C \setminus \Sigma$ is a biholomorphism.

Proof. As before, we start by constructing a resolution for a small neighborhood of a singular point.

Let $0 \in C \subset \mathbb{C}^n$ be a 1-dimensional complex analytic singularity. That is, $0 \in C$ is reduced, and there are holomorphic functions $f_1, \ldots, f_k$ such that

$$C \cap B_\epsilon = (f_1 = \cdots = f_k = 0) \cap B_\epsilon,$$

where $B_\epsilon \subset \mathbb{C}^n$ denotes the ball of radius ϵ around the origin.

A general hyperplane $0 \in H \subset \mathbb{C}^n$ intersects C in a discrete set of points, and hence by shrinking ϵ we may assume that $C \cap H \cap B_\epsilon = \{0\}$.

Let $\pi : \mathbb{C}^n \to \mathbb{C}$ denote the projection with kernel H, and choose coordinates $x_1, \ldots, x_n$ such that π is the nth coordinate projection.

By the implicit function theorem, the set of points where $\pi : C \to \mathbb{C}$ is not a local analytic biholomorphism is given by the condition

$$\operatorname{rank}\left(\frac{\partial f_i}{\partial x_j} : 1 \le i \le k, 1 \le j \le n-1\right) \le n-2.$$

It is thus a complex analytic subset of C. By Sard's theorem it has measure zero and, hence, is a discrete set.

Thus by choosing η small enough, we can assume that

$$\pi : C \cap B_\epsilon \cap \pi^{-1}(\Delta_\eta) \to \Delta_\eta$$

is proper and a local analytic biholomorphism except at the origin, where $\Delta_\eta \subset \mathbb{C}$ is the disc of radius η. Set

$$C_\eta := C \cap B_\epsilon \cap \pi^{-1}(\Delta_\eta) \quad \text{and} \quad C_\eta^* := C_\eta \setminus \{0\}.$$

We thus conclude that

$$\pi : C_\eta^* \to \Delta_\eta^* \quad \text{is a covering map.}$$

The rest of the proof now goes the same as before.   □

1.3. The Albanese method using projections

In algebraic geometry, the simplest method to resolve singularities of curves was discovered by Albanese [Alb24a]. This relies on comparing the singularities of a curve with the singularities of its projection from a singular point.

In order to get a feeling for this, let us consider some examples.

EXAMPLE 1.9. (1) Let $p \in \mathbb{P}^n$ be a point and $\mathbb{P}^{n-1} \cong H \subset \mathbb{P}^n$ a hyperplane not containing p. The *projection* $\pi_{p,H} : \mathbb{P}^n \dashrightarrow H$ of $\mathbb{P}^n$ from p to H is defined as follows. Pick any point $q \ne p$. Then the line through p, q intersects H in the image point $\pi_{p,H}(q)$.

We can choose coordinates on $\mathbb{P}^n$ such that $p = (0:0:\cdots:0:1)$ and $H = (x_n = 0)$. Then

$$\pi_{p,H}(x_0 : \cdots : x_n) = (x_0 : \cdots : x_{n-1}).$$

(2) Assume that $p \in C \subset \mathbb{P}^n$ is a singular point, where m smooth branches of the curve pass through with different tangent directions. Projecting C from p separates these tangent directions, and the singular point p is replaced by m points with only one local branch through each of them.

(3) Let $C \subset \mathbb{A}^n$ be given by the monomials $t \mapsto (t^{m_1}, t^{m_2}, \ldots, t^{m_n})$. We assume that $m_1 < m_i$ for $i \geq 2$.

Projecting from the origin to $\mathbb{P}^{n-1}$, the origin is replaced by a single point $(1 : 0 : \cdots : 0)$, and in the natural affine coordinates we get the parametric curve $t \mapsto (t^{m_2 - m_1}, \ldots, t^{m_n - m_1})$.

This is quite curious. The new monomial curve looks simpler than the one we started with, but it is hard to pin down in what way. For instance, its multiplicity $\min\{m_i - m_1\}$ may be bigger than the multiplicity of the original curve, which is m_1.

(4) Projection may also create singular points. If a line through p intersects C in two or more points or is tangent to C in one point, we get new singular points after projecting.

For $n \geq 4$ these do not occur when p is in general position, but the singular points of C are not in general position, so it is hard to determine what exactly happens. The best one could hope for is that such projections do not create new singular points for general embeddings $C \hookrightarrow \mathbb{P}^n$.

It is quite surprising that for curves of low degree in $\mathbb{P}^n$ we do not have to worry about general position or about ways of measuring the improvement of singularities step-by-step. The intermediate stages may get worse, but the process takes care of itself in the end.

ALGORITHM 1.10 (Albanese). Let $C_0 \subset \mathbb{P}^n$ be a projective curve. If $C_i \subset \mathbb{P}^{n-i}$ is already defined, then pick any singular point $p_i \in C_i \subset \mathbb{P}^{n-i}$ and set

$$C_{i+1} := \overline{\pi_i(C_i)},$$

where $\pi_i : \mathbb{P}^{n-i} \dashrightarrow \mathbb{P}^{n-i-1}$ is the projection from the point p_i.

THEOREM 1.11 (Albanese, 1924). *Let $C_0 \subset \mathbb{P}^n$ be an irreducible, reduced projective curve spanning $\mathbb{P}^n$ over an algebraically closed field.*

If $\deg C_0 < 2n$, then the Albanese algorithm eventually stops with a smooth projective curve $C_m \subset \mathbb{P}^{n-m}$, which is birational to C_0.

COROLLARY 1.12. *Every irreducible, reduced projective curve C over an algebraically closed field can be embedded into some $\mathbb{P}^n$ such that $\deg C < 2n$ and C spans $\mathbb{P}^n$.*

Thus the Albanese algorithm eventually stops with a smooth projective curve $C_m \subset \mathbb{P}^{n-m}$, which is birational to C. The inverse map $C_m \dashrightarrow C$ is a morphism, and thus $C_m \to C$ is a resolution of C.

Proof. All we need is to find a very ample line bundle L on C such that

$$\deg L < 2(h^0(C, L) - 1) \quad \text{or, equivalently,} \quad h^0(C, L) > \tfrac{1}{2} \deg L + 1.$$

Let us see first how to achieve this using the Riemann-Roch theorem for singular curves (which is way too advanced for such an elementary consequence).

The Riemann-Roch theorem says that if L is any line bundle on an irreducible, reduced projective curve, then

$$h^0(C, L) - h^1(C, L) = \deg L + 1 - p_a(C),$$

where the arithmetic genus $p_a(C)$ is easiest to define as $p_a(C) := h^1(C, \mathcal{O}_C)$.

Thus any very ample line bundle L of degree $\geq 2p_a(C) + 1$ works. □

For those who prefer a truly elementary proof of resolution for curves, here is a method to find the required line bundles.

1.13 (Very weak Riemann-Roch on curves). We claim that for any very ample line bundle L,

$$h^0(C, L^m) \geq m \deg L + 1 - \binom{\deg L - 1}{2} \quad \text{for } m \geq \deg L.$$

Indeed, embed C into P^n by L, and then project it generically to a plane curve of degree $\deg L$, $\pi : C \to C' \subset \mathbb{P}^2$. Now, for $m \geq \deg L$,

$$
\begin{aligned}
h^0(C, L^m) &\geq h^0\big(C', \mathcal{O}_{\mathbb{P}^2}(m)|_{C'}\big) \\
&\geq h^0\big(\mathbb{P}^2, \mathcal{O}_{\mathbb{P}^2}(m)\big) - h^0\big(\mathbb{P}^2, \mathcal{O}_{\mathbb{P}^2}(m - \deg L)\big) \\
&= m \deg L + 1 - \binom{\deg L - 1}{2}.
\end{aligned}
$$

Taking any $m \geq \deg L$ is sufficient for the Albanese method. □

Before we start the proof of (1.11), we need some elementary lemmas about space curves and their projections.

LEMMA 1.14. Let $C \subset \mathbb{P}^n$ be an irreducible and reduced curve, not contained in any hyperplane. Then $\deg C \geq n$. Furthermore, if $p_1, \ldots, p_n$ are n distinct points of C, then $\deg C \geq \sum_i \operatorname{mult}_{p_i} C$.

Proof. Pick n points $p_1, \ldots, p_n \in C$, and let $L \subset \mathbb{P}^n$ be the linear span of these points. Then $\dim L \leq n - 1$.

By assumption C is not contained in L, and thus there is a hyperplane $H \subset \mathbb{P}^n$ containing L but not containing C. Thus the intersection $H \cap C$ is finite, and it contains at least n points. This implies that $\deg C \geq n$.

If some of the p_i are singular, then we can further improve the estimate to $\deg C \geq \sum_i \operatorname{mult}_{p_i} C$, (cf. (1.20)). □

1.15 (Projections of curves). Let $A \subset \mathbb{P}^n$ be any irreducible, reduced curve and $p \in A$ a point. Let $\pi : \mathbb{P}^n \dashrightarrow \mathbb{P}^{n-1}$ denote the projection from p as in (1.9).

π is not a morphism, but it becomes one after blowing up p. Let $A' \subset B_0\mathbb{P}^n$ be the birational transform of A and let $A'_p \subset A'$ denote the preimage of p. The closure A_1 of the projection of A is the union of $\pi(A \setminus \{p\})$ and of the image of A'_p.

Claim 1.15.1. Let the notation be as above.

(i) If A spans $\mathbb{P}^n$, then A_1 spans $\mathbb{P}^{n-1}$.

(ii) If $A \dashrightarrow A_1$ is birational, then

$$\deg A_1 = \deg A - \operatorname{mult}_p A.$$

(iii) If $A \dashrightarrow A_1$ is not birational, then

$$\deg A_1 \cdot \deg(A/A_1) = \deg A - \operatorname{mult}_p A.$$

Proof. If $A_1 \subset H$ for some hyperplane H, then $A \subset \pi^{-1}(H)$, a contradiction, proving (i).

In order to see (ii), let $H \subset \mathbb{P}^{n-1}$ be a general hyperplane. It intersects A_1 in $\deg A_1$ points. If H avoids all the points over which $\pi : A \setminus \{p\} \to A_1$ is not a local isomorphism, then $\pi^{-1}(H)$ intersects A in the points $H \cap A_1$ and also at p. Here $\pi^{-1}(H)$ is a general hyperplane through p thus the intersection number of A and $\pi^{-1}(H)$ at p is $\operatorname{mult}_p A$ by (1.20).

Finally, the same argument as before shows (iii), once we notice that $\deg(A/A_1)$ points of A lie over a general point of A_1. We have to be a little more careful when $A \to A_1$ is inseparable, when the total number of preimages is the degree divided by the degree of inseparability. However, the local intersection multiplicity goes up by the degree of inseparability (1.20), and so the two changes cancel each other out.   □

1.16 (Proof of (1.11)). Starting with $C_0 \subset \mathbb{P}^n$ such that $\deg C_0 < 2n$, we get a sequence of curves $C_i \subset \mathbb{P}^{n-i}$.

If the projection $C_{i-1} \dashrightarrow C_i$ is birational, then $\deg C_i \le \deg C_{i-1} - 2$ by (1.15.1), and thus $\deg C_i < 2(n - i)$.

If $C_i \dashrightarrow C_{i+1}$ is not birational and we project from a point of multiplicity $m_i \ge 2$, then

$$\deg C_{i+1} = \frac{\deg C_i - m_i}{\deg(C_i/C_{i+1})} \le \frac{\deg C_i - m_i}{2} < n - i - 1.$$

On the other hand, $\deg C_{i+1} \ge n - i - 1$ by (1.14), a contradiction.

Thus all the projections are birational. The sequence of projections must stop after at most $n - 1$ steps, so at some stage we get $C_m \subset \mathbb{P}^{n-m}$ without any points of multiplicity at least 2. Therefore, C_m is smooth.   □

1.17 (The Albanese method over nonclosed fields). Let C be an irreducible and reduced curve over a field k, which is not algebraically closed, and choose an embedding $C \subset \mathbb{P}^n$ such that $\deg C < 2n$.

If $p \in C(k)$ is a singular point defined over k, then we can proceed without any change and project from p as before.

What happens with singular points $p \in C(\bar{k})$ that are defined over an extension field $k' \supset k$?

If k'/k is separable (for instance, if $\operatorname{char} k = 0$), then the point p has several conjugates $p = p_1, \ldots, p_d$ and the linear space $L = \langle p_1, \ldots, p_d \rangle \subset \mathbb{P}^n$ spanned by them is defined over k. Thus we can project from L to get

$\pi_L : \mathbb{P}^n \dashrightarrow \mathbb{P}^{n-d}$, and everything works out as before. (Note that $L \neq \mathbb{P}^n$ by (1.14).)

(As long as the p_i are in general position, over $\bar{k}$ we can also realize π_L as projecting from the points $p_1, \ldots, p_d$ successively.)

From this we conclude the following.

COROLLARY 1.18. *Let k be a perfect field and C an irreducible, reduced projective curve over k.*

Then C can be embedded into some $\mathbb{P}^n$ such that $\deg C < 2n$ and the Albanese algorithm eventually stops with a smooth projective curve, which is a resolution of C. □

1.19 (Curves over nonperfect fields). The typical example of nonperfect fields is a one-variable function field $K = k(t)$, where k is any field of characteristic p.

Over K consider the hyperelliptic curve $C := (y^2 = x^p - t)$. After adjoining $t^{1/p}$ this can be rewritten as $(y^2 = (x - t^{1/p})^p)$, which is singular at the point $(t^{1/p}, 0)$.

Nonetheless, the original curve C is nonsingular; that is, its local rings are regular. Indeed, the only point in question is $(t^{1/p}, 0)$, and over k it is defined by the equations $y = x^p - t = 0$. The maximal ideal of this point in $K[x, y]/(y^2 - x^p + t)$ is thus $(y, x^p - t)/(y^2 - x^p + t)$, which is generated by y alone.

Even worse examples appear over a two-variable function field $K = k(s, t)$. I leave it to the reader to check that $(sx^p + ty^p + z^p = 0) \subset \mathbb{P}^2$ is nonsingular over K, but over the algebraic closure $\bar{K}$ it becomes the p-fold line since

$$sx^p + ty^p + z^p = (s^{1/p}x + t^{1/p}y + z)^p.$$

Curves like this certainly cannot be made smooth by projections since they are not even birational to any smooth projective curve. Of course here the relevant question is, does the Albanese algorithm produce the nonsingular model of C over nonperfect fields?

The basic inductive structure of the proof breaks down in some examples, and I am not sure what happens in general.

1.20 (Review of multiplicities, I). Almost every introduction to algebraic geometry discusses the order of zero or pole of a rational function on a smooth curve, but very few consider these notions for singular curves. Here we give a short sketch of the general case.

Let k be a field, C an irreducible and reduced curve defined over k and $P = \{p_1, \ldots, p_n\} \subset C$ a finite number of closed points. Let $R = \mathcal{O}_{P,C}$ be the semi-local ring of P; that is, we invert every function that is nonzero at all the p_i. Then R is a 1-dimensional integral domain with finitely many maximal ideals $m_i = m_{p_i}$ and $\dim_k R/m_i < \infty$. For us the best definition

of 1-dimensional is that $\dim_k R/(r) < \infty$ for every $r \in R^* := R \setminus \{0\}$. Let $K \supset R$ denote the quotient field with multiplicative group $K^* := K \setminus \{0\}$.

More generally, everything works if R is a 1-dimensional semi-local ring without nilpotent elements, $R^* \subset R$ is the group of non–zero divisors, $K \supset R$ is obtained by inverting R^* and $K^* \subset K$ denotes the subgroup of invertible elements.

If $R = k[x]_{(x)}$ and $f = x^m(\text{unit})$, then $m = \dim_k R/fR$ is the usual order of vanishing or multiplicity of f. Based on this, for any R and any $f \in R^*$ we call $e_R(f) := \dim_k R/fR$ the multiplicity of f.

If R/m is bigger than k, then one may think that this is the "usual" order of vanishing times $\dim_k R/m$. A typical example is $C = (y^2 = x^2 - 1) \subset \mathbb{A}^2_\mathbb{R}$ and P corresponds to the maximal ideal (x). We compute that $\dim_\mathbb{R} \mathcal{O}_{P,C}/(x) = 2$. Looking at it over $\mathbb{C}$, we see that x vanishes at two points $(0, \pm i)$, with multiplicity 1 at each. Thus the total order of vanishing is 2, and our definition works well after all.

It is convenient to think of a function with a pole as having a negative order of vanishing. With this in mind, we define the multiplicity of any $f \in K^*$ as follows. Write $f = r_1/r_2$, where $r_i \in R^*$. Thus $R/r_1 R$ and $R/r_2 R$ are both finite dimensional. Set

$$e_P(f) := e_R(f) := \dim_k R/r_1 R - \dim_k R/r_2 R, \qquad (1.20.1)$$

and call it the *multiplicity* of f at p. If we write $f = (r_1 r_3)/(r_2 r_3)$, then

$$\dim_k R/r_1 r_3 R - \dim_k R/r_2 r_3 R$$
$$= \dim_k R/r_1 R + \dim_k r_1 R/r_1 r_3 R - \dim_k R/r_2 R - \dim_k r_2 R/r_2 r_3 R$$
$$= \dim_k R/r_1 R - \dim_k R/r_2 R,$$

where the last equality holds since multiplication by r_2/r_1 gives an isomorphism $r_1 R/r_1 r_3 R \to r_2 R/r_2 r_3 R$. So the notion is well defined. As we change field from k to $k' \supset k$, the k-vector space $R/r_i R$ is replaced by a k'-vector space of the same dimension (namely, $k' \otimes_k R/r_i R$), so the multiplicity does not depend on k.

Note however that $e_P(f) = 0$ does not imply that f is a unit at P, not even if P consists of a single point. For instance, $f = x/y$ has multiplicity 0 at the origin of the irreducible cubic $(xy + x^3 + y^3 = 0)$.

The multiplicity is additive; that is,

$$e_P(fg) = e_P(f) + e_P(g). \qquad (1.20.2)$$

Indeed, write $f = r_1/r_2$ and $g = s_1/s_2$. Then

$$e_P(fg) = \dim_k R/r_1 s_1 R - \dim_k R/r_2 s_2 R$$
$$= \dim_k R/r_1 R + \dim_k r_1 R/r_1 s_1 R - \dim_k R/r_2 R - \dim_k r_2 R/r_2 s_2 R$$
$$= \dim_k R/r_1 R + \dim_k R/s_1 R - \dim_k R/r_2 R - \dim_k R/s_2 R$$
$$= e_P(f) + e_P(g).$$

Another important property of multiplicity is semi-continuity.

Lemma 1.20.3. For $f_1, \ldots, f_n \in K^*$, the (partially defined) function $k^n \to \mathbb{Z}$ given by

$$\sum t_i f_i \to e_R\left(\sum t_i f_i\right)$$

takes its minimum on an open subset of k^n.

Proof. Multiply through the common denominator to assume that $f_i \in R$. Fix $r \in m_R \cap R^*$. For every natural number s,

$$
\begin{aligned}
e_R\left(\sum t_i f_i\right) &= \dim_k R/\left(\sum t_i r_i\right)R \geq \dim_k R/\left(r^s, \sum t_i r_i\right)R \\
&= \dim_k R/r^s R - \operatorname{rank}_k\left[\sum t_i r_i : R/r^s R \to R/r^s R\right].
\end{aligned}
$$

The rank is a lower-semi-continuous function for families of k-linear maps of finite-dimensional k-vector spaces. We are done if we can prove that there is a fixed s such that $r^s \in \left(\sum t_i r_i\right)R$ for every $(t_1, \ldots, t_n)$ in a suitable dense open subset of k^n.

To show this, introduce new independent variables T_i, and work over the field $k' = k(T_1, \ldots, T_n)$. The new ring $R_{k'}$ is still 1-dimensional and local, and so $r^s \in \left(\sum T_i r_i\right)R_{k'}$ for some s. This means that

$$
r^s = \left(\sum T_i r_i\right) \cdot \sum_j \frac{\phi_j(T_1, \ldots, T_n)}{\psi_j(T_1, \ldots, T_n)} \cdot b_j
$$

for some $b_j \in R$ and $\phi_j, \psi_j \in k[T_1, \ldots, T_n]$. For $t_1, \ldots, t_n$ in the open set $\prod_j \psi_j \neq 0$ we can substitute $T_i = t_i$ to obtain $r^s \in \left(\sum t_i r_i\right)R$. $\square$

When $P = p$ is a single point and k is an infinite field, we define the *multiplicity* of $p \in C$ by

$$
\operatorname{mult}_p C := \frac{1}{\dim_k \mathcal{O}_{p,C}/m_p} \min\{e_p(f) : f \in m_p\}. \tag{1.20.4}
$$

Thus $\operatorname{mult}_p C = 1$ iff $m_p \subset \mathcal{O}_{p,C}$ is a principal ideal.

Assume for example that $C \subset \mathbb{A}^2$ is defined by an equation $(g = 0)$ and $p \in C$ is the origin. We show that $\operatorname{mult}_p C$ coincides with the multiplicity of g at the origin, that is, the degree of the lowest monomial in g.

If $\operatorname{mult}_p g = m$, then $g \in (x, y)^m$, and hence we get a surjection $\mathcal{O}_C \twoheadrightarrow k[x, y]/(x, y)^m$. We can assume that $f = y + (\text{other terms})$ and then

$$
e_p(f) \geq \dim_k k[x, y]/\left(f, (x, y)^m\right) = \dim_k k[x]/(x^m) = m.
$$

Conversely, choose a general linear function for f. After a coordinate change we can assume that $f = y$, and the generic choice assures that x^m appears in g with nonzero coefficient. Thus we can write $g = yu + x^m v$, where $v(p) \neq 0$ and set $L := (y = 0)$. Then

$$
\mathcal{O}_{p,C}/(f) = \mathcal{O}_{p,\mathbb{A}^2}/(g, y) = \mathcal{O}_{p,L}/(g) = \mathcal{O}_{p,L}/(x^m v) \cong k[x]/(x^m).
$$

This shows that $e_p(f) \leq m$; thus in fact $e_p(f) = m$.

If k is finite, a dense open subset of $\mathbb{A}_k^n$ need not have any k-points, so the above definition does not work. See Section 2.9 for a general definition.

Lemma 1.20.5. Let C be a reduced, irreducible projective curve. Then $\sum_{p \in C} e_P(f) = 0$ for every nonzero rational function $f \in k(C)$.

Proof. If C is smooth, then $f : C \to \mathbb{P}^1$ is everywhere defined. If C is singular, then we only get a rational map $f : C \dashrightarrow \mathbb{P}^1$ and this causes problems. Our first step is to reduce to the case where $f : C \to \mathbb{P}^1$ is a morphism.

To do this, let h be a rational function on C, which is contained in the local ring of every singular point. Then fh^s is also in the local ring of every singular point for $s \gg 1$. Writing $f = (fh^s)/h^s$ and using additivity (1.20.2), it is enough to prove the assertion for those functions f that are contained in the local ring of every singular point. Such an f can be viewed as a finite morphism $f : C \to \mathbb{P}^1$. Let $C_0 := C \setminus (\text{polar set of } f)$ and $C_\infty := C \setminus (\text{zero set of } f)$. Then $k[C_0]$ is a finite and torsion-free $k[f]$-module, and hence free. Similarly, $k[C_\infty]$ is a free $k[f^{-1}]$-module. Thus

$$\begin{aligned}
\sum_{p \in C_0} e_p(f) &= \dim_k k[C_0]/fk[C_0] = \mathrm{rank}_{k[f]} k[C_0], \quad \text{and} \\
-\sum_{p \in C_\infty} e_p(f) &= \dim_k k[C_\infty]/f^{-1}k[C_\infty] = \mathrm{rank}_{k[f^{-1}]} k[C_\infty].
\end{aligned}$$

Since $\mathrm{rank}_{k[f]} k[C_0] = \mathrm{rank}_{k[f,f^{-1}]} k[C_0 \cap C_\infty] = \mathrm{rank}_{k[f^{-1}]} k[C_\infty]$, we are done. $\qquad\square$

Let L be a line bundle on C and s_1 any nonzero rational section of L. At each point $p \in C$ we can identify the $\mathcal{O}_{p,C}$-module L with $\mathcal{O}_{p,C}$ and define the multiplicity $e_p(s_1)$. The *degree* of L on C is defined by the formula

$$\deg_C L := \sum_p e_p(s_1). \tag{1.20.6}$$

If s_2 is another section, then by (1.20.2) and (1.20.3) we get that

$$\sum_p e_p(s_1) - \sum_p e_p(s_2) = \sum_p e_p(s_1/s_2) = 0,$$

so $\deg_C L$ is well defined and does not depend on k.

From the definition we see that if L has a nonzero section, then $\deg_C L \geq 0$. If s is a nonzero section of L and $p \in C$ a k-point such that $s(p) \neq 0$ then in passing from L to $L(-p)$ we lose one section and lower the degree by 1. If k is algebraically closed, then we have plenty of points such that $s(p) \neq 0$. Repeating this if necessary we get the basic inequality between degrees and the space of global sections:

$$\dim_k H^0(C, L) \leq \deg_C L + 1 \quad \text{if } \deg_C L \geq 0. \tag{1.20.7}$$

This holds for arbitrary k as well, since both sides remain unchanged under field extensions. (For the left-hand side, this is explained in [Sha94, III.3.5].)

Remark 1.20.8. Lemma (1.20.5) also holds for invertible rational functions if C is 1-dimensional and projective, with no embedded points. This can be used to define the degree of any line bundle. The inequality (1.20.7), however, no longer holds in general, since even $\mathcal{O}_C$ can have many sections.

1.4. Normalization using commutative algebra

These days, commutative algebra is part of the familiar foundations of algebraic geometry, and it is hard to imagine that the relationship between resolutions and normalization was not always obvious. So we start this section by trying to explain how one may be led from complex analytic resolutions to normalization. Then we prove that normalization does give the resolution for 1-dimensional integral domains that are finitely generated over a field.

1.21 (Why normalization?). Let C be a complex analytic curve with resolution $n : \bar{C} \to C$. Let $\Sigma \subset C$ denote the set of singular points and $C^0 \subset C$ the open subset of smooth points. Let $\bar{\Sigma} = n^{-1}(\Sigma)$ and $\bar{C}^0 = n^{-1}(C^0)$ denote their preimages. Then $n : \bar{C}^0 \to C^0$ is an isomorphism.

By restriction we obtain the maps below, where the first one is an isomorphism by the Riemann extension theorem and the second is an isomorphism since $\bar{C}^0 \cong C^0$:

$$\left\{ \begin{array}{c} \text{holomorphic} \\ \text{functions on } \bar{C} \end{array} \right\} \cong \left\{ \begin{array}{c} \text{holomorphic} \\ \text{functions on } \bar{C}^0 \\ \text{bounded near } \bar{\Sigma} \end{array} \right\} \cong \left\{ \begin{array}{c} \text{holomorphic} \\ \text{functions on } C^0 \\ \text{bounded near } \Sigma \end{array} \right\}.$$

Looking at the right-hand side, we obtain an intrinsic way of defining the ring (or sheaf) of holomorphic functions on $\bar{C}$ as the ring (or sheaf) of holomorphic functions on C^0 that are bounded near the singular points.

It is quite natural to expect that a similar description would hold in the algebraic setting as well.

Thus let C be a complex, affine algebraic curve with singular set Σ and let $C^0 := C \setminus \Sigma$, be the set of smooth points. In analogy with the holomorphic case, one arrives at the conjecture

$$\left\{ \begin{array}{c} \text{regular} \\ \text{functions on } \bar{C} \end{array} \right\} \overset{?}{\cong} \left\{ \begin{array}{c} \text{regular} \\ \text{functions on } C^0 \\ \text{bounded near } \Sigma \end{array} \right\}.$$

This looks pretty good, but boundedness is not really a concept of algebraic geometry. To understand it, let us look again on $\bar{C}$.

If $n : \bar{C} \to C$ is a resolution, then n is birational; thus every (regular or rational) function on $\bar{C}$ is also a rational function on C.

Let f be a regular function on $\bar{C}^0$ that is not bounded along $\bar{\Sigma}$. Since f is a rational function on $\bar{C}$, it must have a pole at one of the points $p \in \bar{\Sigma}$. Thus by looking at Laurent series around p, we get a homomorphism

$$\Phi_\Sigma : \left\{ \begin{array}{c} \text{rational} \\ \text{functions on } C \end{array} \right\} \to \sum_{p \in \bar{\Sigma}} \left\{ \begin{array}{c} \text{Laurent series} \\ \text{around } p \end{array} \right\},$$

which has the property that a rational function f is bounded around Σ iff its Laurent series expansion has no pole for every $p \in \bar{\Sigma}$.

For algebraic varieties one could keep working with Laurent series, but in general it is more convenient to change to arbitrary DVRs.

DEFINITION 1.22. A *discrete valuation ring* or *DVR* is a Noetherian integral domain R with a single maximal ideal that is also principal; that is, $m = (t)$ for some $t \in R$.

This easily implies (cf. [AM69, 9.2]) that every element of the quotient field $Q(R)$ can be written uniquely as $t^n u$, where $n \in \mathbb{Z}$ and $u \in R \setminus m$ is a unit. Hence, just as in Laurent series fields, we can talk about an element having a pole of order $-n$ (if $n < 0$) or a zero of order n (if $n > 0$).

We can now state the first definition of normality, which is inspired by the above considerations.

DEFINITION 1.23. Let S be an integral domain with quotient field $Q(S)$. The *normalization* of S in $Q(S)$, denoted by $\bar{S}$, is the unique largest subring $\bar{S} \subset Q(S)$ such that every homomorphism $\phi : S \to R$ to a DVR extends to a homomorphism $\bar{\phi} : \bar{S} \to R$.

Note that ϕ has a unique extension to a partially defined homomorphism $\Phi : Q(S) \dashrightarrow Q(R)$ between the quotient fields given by

$$\Phi(s_1/s_2) := \phi(s_1)/\phi(s_2), \quad \text{whenever } \phi(s_2) \neq 0.$$

Thus $\bar{S} = \cap_{\phi:S \to R} \Phi^{-1}(R)$.

In general, it is quite difficult to use this definition to construct the normalization, but here is a useful case that is easy.

LEMMA 1.24. *A unique factorization domain is normal. In particular, any polynomial ring* $k[x_1, \ldots, x_n]$ *over a field is normal.*

Proof. If $p_i \in S$ are the irreducible elements, then every element of $Q(S)$ can be uniquely written as a finite product $u \prod_i p_i^{m_i}$, where $u \in S$ is a unit and $m_i \in \mathbb{Z}$.

For every p_j we define

$$S_{p_j} := \{u \prod_i p_i^{m_i} : m_j \geq 0\} \subset Q(S).$$

Then S_{p_j} is a DVR whose maximal ideal is generated by p_j and $S = \cap_{p_j} S_{p_j}$. □

The following observation leads to a quite different definition of normality, which comes more from the study of algebraic number fields.

LEMMA 1.25. *Let the notation be as in (1.23). Assume that $t \in Q(S)$ satisfies a monic equation*

$$t^m + s_{m-1} t^{m-1} + \cdots + s_0 = 0, \quad \text{where } s_i \in S.$$

Then $t \in \bar{S}$.

Proof. Pick any $\phi : S \to R$, and consider its extension Φ. We get that

$$\Phi(t)^m + \phi(s_{m-1})\Phi(t)^{m-1} + \cdots + \phi(s_0) = 0.$$

If $\Phi(t)$ has a pole of order $r > 0$, then $\Phi(t)^m$ has a pole of order mr, while all the other terms of the equation have a pole of order at most $(m-1)r$. This is impossible, and hence $\Phi(t) \in R$. □

DEFINITION 1.26. Let $S \subset S'$ be a ring extension. We say that $s \in S'$ is *integral* over S if one of the following equivalent conditions holds.

(1) s satisfies a monic equation

$$s^m + r_{m-1} s^{m-1} + \cdots + r_0 = 0, \quad \text{where } r_i \in S.$$

(2) The subring $S[s] \subset S'$ is a finitely generated S-module.

It is easy to see that all elements integral over S form a subring, called the *integral closure* or *normalization* of S in S'.

DEFINITION 1.27. Let S be an integral domain. The *normalization* of S is its integral closure in its quotient field $Q(S)$.

This is now the standard definition in commutative algebra books. See, for instance, [AM69, Chap.5.Sec.1] for the basic properties that we use.

REMARK 1.28. The fact that these definitions (1.23) and (1.27) are the same is not obvious. One implication is given by (1.25).

The easy argument that every normal integral domain S is the intersection of all the valuation rings V sitting between it and its quotient field $S \subset V \subset Q(S)$ is in [AM69, 5.22]. Working only with discrete valuation rings is a bit harder. The strongest theorem in this direction is Serre's criterion for normality; see [Mat70, 17.I] or [Mat89, 23.8]. We do not need it for now.

1.29 (Is normalization useful?). With the concept of normalization established, we have to see how useful it is. In connection with resolutions, two questions come to mind.

(1) If C is an affine curve whose coordinate ring $k[C]$ is normal, does that imply that C is smooth?

(2) Does the normalization give the resolution?

The answer to the first question is yes, as we see it in (1.30). This of course begs another question, how do we see smoothness in terms of commutative algebra?

Just look at a plane curve C with coordinate ring $k[x,y]/(f(x,y))$ and let $m \subset k[x,y]/(f(x,y))$ be the ideal of the origin. The following are easily seen to be equivalent.

(i) C is smooth at the origin.

(ii) $f(0,0) = 0$ and f contains a linear term.

(iii) $\dim_k m/m^2 = 1$.

In general, we say that a 1-dimensional ring is *regular* if for any maximal ideal $m \subset R$ the quotient m/m^2 is 1-dimensional as an R/m-vector space. We end up proving that every 1-dimensional normal ring is regular. (The converse is easy.)

The second question is more troublesome. When constructing resolutions, we need $f : \bar{C} \to C$ to be surjective. By the going-up and -down theorems (cf. [AM69, 5.11, 5.16]) we are in good shape if $k[\bar{C}]$ is *finite* over $k[C]$. (That is, $k[\bar{C}]$ is finitely generated as a module over $k[C]$.) This is nowadays adopted as a necessary condition of resolution. For coordinate rings of algebraic varieties, the normalization is finite (1.33).

It should be noted though that finiteness of normalization fails in some examples (cf. (1.103) or (1.105)), but one could reasonably claim that the normalization is nonetheless the resolution.

The following is the nicest result relating normalization to resolutions.

THEOREM 1.30. *Let R be a 1-dimensional, normal, Noetherian integral domain. Then R is regular. That is, for every maximal ideal $m \subset R$, the quotient m/m^2 is 1-dimensional (over R/m).*

1.31 (Nuts-and-bolts proof). For C a singular, reduced and affine curve over an algebraically closed field k, we construct an explicit rational function in $k(C) \setminus k[C]$ that is integral over $k[C]$. Besides proving (1.30), this also gives an algorithm to construct the normalization.

Pick $p \in C$, and assume that m_p/m_p^2 is at least 2-dimensional. Pick any $x, y \in m_p$ that are independent in m_p/m_p^2.

Claim 1.31.1. There are $a \in k$ and $u \in k[C] \setminus m_p$ such that

$$\frac{y}{x + ay} u \notin k[C] \quad \text{is integral over } k[C].$$

Proof. An element of this form is definitely not in $k[C]$. Indeed, otherwise we would have that

$$uy = (x + ay)\frac{y}{x + ay} u \in (x + ay)k[C],$$

so uy would be a multiple of $x + ay$ in m_p/m_p^2, a contradiction.

x, y can be viewed as elements of the function field $k(C)$, which has transcendence degree 1 over k. Thus x, y satisfy an algebraic relation $f(x, y) = 0$. Let $f_m(x, y)$ be the lowest degree homogeneous part.

Make a substitution $x \mapsto x + ay$. The coefficient of y^m in $f_m(x, y)$ is now $f_m(a, 1)$. We can choose a different from zero such that after multiplying by a suitable constant we achieve that

$$f(x, y) = y^m + \left(\text{other terms } (x + ay)^i y^j \text{ of degree at least } m \right).$$

We can rewrite this as

$$y^m \left(1 + g_m(x, y) \right) + y^{m-1}(x + ay)g_{m-1}(x, y) + \cdots + (x + ay)^m g_0(x, y) = 0,$$

where $g_m(x, y) \in (x, y) \subset m$. Hence $u := 1 + g_m(x, y)$ is not in m_p and $g_i \in k[C]$ for $i < m$. Thus over $k[C]$ we get the integral dependence relation

$$\left(\frac{y}{x + ay} u \right)^m + \left(\frac{y}{x + ay} u \right)^{m-1} \left(g_{m-1}(x, y) \right) + \cdots + \left(u^{m-1} g_0(x, y) \right) = 0. \quad \square$$

1.32 (Slick proof). By localizing at m_p we are reduced to the case where (R, m) is local. Pick an element $x \in m \setminus m^2$. If $m = (x)$, then we are done. Otherwise, $m/(x)$ is nonzero. Since R has dimension 1, $R/(x)$ is 0-dimensional, so $m/(x)$ is killed by a power of m. Thus there is a $y \in m \setminus (x)$ such that $my \in (x)$. Equivalently,

$$\frac{y}{x} m \subset R.$$

If $\frac{y}{x} m$ contains a unit, then $\frac{y}{x} z = u$ for some $z \in m$ and a unit u; thus $x = yzu^{-1} \in m^2$, which is impossible.

Thus $\frac{y}{x} m \subset m$. Now we can run the beautiful proof of the Nakayama lemma (cf. [AM69, 2.4]), which is worth repeating.

Write $m = (x_1, \dots, x_n)$. Then there are $r_{ij} \in R$ such that

$$\frac{y}{x} x_j = \sum_i r_{ij} x_j.$$

Thus the vector $(x_1, \dots, x_n)$ is a null vector of the matrix

$$\frac{y}{x} \mathbf{1}_n - (r_{ij}),$$

and hence its determinant is zero. This determinant is a monic polynomial in $\frac{y}{x}$ with coefficients in R, and hence $\frac{y}{x} \in R$ since R is normal.

This, however, means that $y \in (x)$, contrary to our choice of y. $\quad \square$

THEOREM 1.33. *Let S be an integral domain that is finitely generated over a field k, and let $F \supset Q(S)$ be a finite field extension of its quotient field. Then the normalization of S in F is finite over S.*

Proof. One's first idea might be to write down any finite extension S'/S whose quotient field is F and reduce everything to the case where $F = Q(S)$. We will see how this works for curves, but in general it seems better to go the other way around, even when $F = Q(S)$ to start with.

By the Noether normalization theorem (1.35), S is finite over a polynomial ring R. So it is enough to prove (1.33) for a polynomial ring $R = k[x_1, \ldots, x_n]$. The key advantage of this reduction is that R is normal.

A short argument proving (1.33) in the case where $F/Q(S)$ is separable and S is normal is in [AM69, 5.17]. Thus we are done in characteristic zero, but in positive characteristic we still have to deal with inseparable extensions.

F is a finite extension of $k(x_1, \ldots, x_n)$, so there is a finite purely inseparable extension $E/k(x_1, \ldots, x_n)$ such that EF/E is separable. Every finite purely inseparable extension of $k(x_1, \ldots, x_n)$ is contained in a field $k'(x_1^{p^{-m}}, \ldots, x_n^{p^{-m}})$, where k'/k is finite and purely inseparable. The normalization of $k[x_1, \ldots, x_n]$ in $k'(x_1^{p^{-m}}, \ldots, x_n^{p^{-m}})$ is $k'[x_1^{p^{-m}}, \ldots, x_n^{p^{-m}}]$ since the latter is a unique factorization domain, hence normal by (1.24).

Thus every finite extension of $k[x_1, \ldots, x_n]$ is contained in a finite and separable extension of $k'[x_1^{p^{-m}}, \ldots, x_n^{p^{-m}}]$, and hence it is finite by the first argument. $\square$

1.34. Here is another approach to (1.33) for projective curves.

Let C be an irreducible, reduced projective curve over an algebraically closed field k. If $C_0 := C$ is not smooth, then as in (1.31) we can write down another curve C_1 and a proper birational map $\pi_1 : C_1 \to C_0$, which is not an isomorphism. We prove that after finitely many iterations we get a smooth curve.

In (1.13) we constructed a line bundle L on C such that

$$h^0(C_0, L^m) \geq m \deg L + 1 - p(C_0, L) \quad \text{for some } p(C_0, L) \geq 0 \text{ and } m \gg 1.$$

Choose that smallest possible value for $p(C_0, L)$. (This is the arithmetic genus, but we do not need to know this for the present argument.) The line bundle $\pi_1^* L^m$ is very ample for $m \gg 1$, so not all of its sections pull back from C_0. Thus

$$h^0(C_1, \pi_1^* L^m) > h^0(C_0, L^m) \quad \text{for } m \gg 1.$$

Hence there is a $p(C_1, L) < p(C_0, L)$ such that

$$h^0(C_1, \pi_1^* L^m) \geq m \deg L + 1 - p(C_1, L) \quad \text{for } m \gg 1.$$

Iterating this procedure, we get curves $\pi_i : C_i \to C_0$ such that

$$h^0(C_i, \pi_i^* L^m) \geq m \deg L + 1 - p(C_0, L) + i \quad \text{for } m \gg 1.$$

Since $h^0(D, M) \leq \deg M + 1$ for any irreducible and reduced curve D and for any line bundle M, we conclude that we get a smooth curve C_j for some $j \leq p(C_0, L)$. $\qquad\square$

In the proofs above we have used the following theorem, due to E. Noether. It is traditionally called the normalization theorem, though it has nothing to do with normalization as defined in this section. Rather, it creates some sort of a "normal form" for all rings that are finitely generated over a field k. This form is not at all unique, but it is very useful in reducing many questions involving general affine varieties to affine spaces.

THEOREM 1.35 (Noether normalization theorem). *Let S be a ring that is finitely generated over a field k. Then there is a polynomial ring $k[x_1, \ldots, x_n] \cong R \subset S$ such that S is finite over R (that is, finitely generated as an R-module).*

Proof. By assumption S can be written as $k[y_1, \ldots, y_m]/(f_1, \ldots, f_s)$.

Look at the first equation f_1, say, of degree d_1. If $y_m^{d_1}$ appears in f_1 with nonzero coefficient, then f_1 shows that y_m is integral over $k[y_1, \ldots, y_{m-1}]$ and we finish by induction.

Otherwise we try to make a change of variables $y_i = y_i' + g_i(y_m)$ for $i < m$ to get a new polynomial equation

$$f_1' := f_1\big(y_1' + g_1(y_m), \ldots, y_{m-1}' + g_{m-1}(y_m), y_m\big),$$

which shows that y_m is integral over $k[y_1', \ldots, y_{m-1}']$.

If k is infinite, then we can get by with a linear change $y_i = y_i' + a_i y_m$. Indeed, after this change the coefficient of $y_m^{d_1}$ in f_1' is $f_{1,d_1}(a_1, \ldots, a_{m-1}, 1)$, where f_{1,d_1} denotes the degree d_1 homogeneous part of f_1. This is nonzero for general $a_1, \ldots, a_{m-1}$.

For finite fields such coordinate changes may not work, and here we use $y_i = y_i' + y_m^{n_i}$.

Assume that f_1 contains a term $x_1^{a_1} \cdots x_m^{a_m}$ with nonzero coefficient, and this is lexicographically the largest one. (That is, a_1 is the largest possible, among those with maximal a_1, then a_2 is the largest, etc.)

As long as the sequence $n_1, \ldots, n_m$ satisfies $n_i > d_1 n_{i+1}$, the highest y_m-power in f_1' is $y_m^{\sum n_i a_i}$, and it occurs with nonzero constant coefficient. So y_m is integral over $k[y_1', \ldots, y_{m-1}']$. $\qquad\square$

1.5. Infinitely near singularities

Let $C \subset \mathbb{C}^2$ be a plane curve given by an equation $f(x, y) = 0$. In order to study the singularity of C at the origin, we write f as a sum of homogeneous terms

$$f(x, y) = f_m(x, y) + f_{m+1}(x, y) + \cdots.$$

The degree of the lowest nonzero term f_m is the *multiplicity* of C at $0 \in \mathbb{C}^2$. It is denoted by $\mathrm{mult}_0\, f$ or by $\mathrm{mult}_0\, C$.

The multiplicity is pretty much the only invariant of the singularity that is easily computable from the equation, but it is not a really good measure of the complexity of the singularity. M. Noether realized that one should view a curve singularity together with all other singularities that can be obtained from it by blowing up. These are the so-called *infinitely near singularities*. The basic observation is that the true complexity of the singularity can be computed from the multiplicities of all infinitely near singularities.

This idea also leads to several nice methods to resolve singularities.

1.36 (Blowing up a smooth surface). Let $0 \in S$ be a point on a smooth surface, and let x, y be local coordinates. Assume for simplicity that x, y do not have any other common zero outside 0. (This can always be achieved by replacing S by a smaller affine neighborhood of $0 \in S$.)

The blow-up of $0 \in S$, denoted by $B_0 S$, is the surface

$$B_0 S := (xu - yv = 0) \subset S \times \mathbb{P}^1_{(u:v)}.$$

Let $\pi : B_0 S \to S$ denote the first coordinate projection. The curve $E = (x = y = 0) \subset B_0 S$, called the *exceptional curve*, is contracted by π to the point 0, and $\pi : B_0 S \setminus E \cong S \setminus \{0\}$ is an isomorphism.

It is usually convenient to work with an affine covering of $B_0 S$. On the $v \neq 0$ chart, we can use $y_1 := u/v$ as a coordinate, and then we have

$$B_0 S_{v \neq 0} = (xy_1 - y = 0) \subset S \times \mathbb{A}^1_{y_1}.$$

Thus if $S = \mathrm{Spec}\, R$, then $B_0 S_{v \neq 0} = \mathrm{Spec}\, R[\frac{y}{x}]$. Similarly, the $u \neq 0$ chart can be represented as $\mathrm{Spec}\, R[\frac{y}{x}]$. From this we see that $B_0 S$ is smooth. See [Sha94, Sec.II.4] for more details.

1.37 (Digression on fields of representatives). If X is an algebraic variety over a field k and $x \in X$ is a point, then we have a corresponding local ring $\mathcal{O}_{x,X}$ with maximal ideal m_x. $\mathcal{O}_{x,X}$ is a k-algebra, but its residue field $\mathcal{O}_{x,X}/m_x$ is usually bigger than k if k is not algebraically closed or if x is not a closed point.

In general $\mathcal{O}_{x,X}$ does not contain any field K that maps isomorphically onto $\mathcal{O}_{x,X}/m_x$. For instance, if $k = \mathbb{C}$ and $x = Y \subset X$ is a subvariety, then the existence of such a field K is equivalent to a rational "retraction" map $X \dashrightarrow Y$, which is the identity on Y.

There is no such map if $X = \mathbb{C}^2$ and if $Y \subset \mathbb{C}^2$ is a nonrational plane curve.

This is somewhat awkward since in any local ring (R, m) it is very convenient to imagine that the elements of m^s/m^{s+1} are degree s homogeneous polynomials in a basis of m/m^2.

Although we do not have any natural section of the quotient map $R \to R/m$, we just pick any set-theoretic lifting $R/m \dashrightarrow R$ (which is neither additive nor multiplicative in general), and this allows us to write things like $\sigma(a)x^i y^j$ for any $a \in R/m$ and $x, y \in m$. This will be unique modulo m^{i+j+1}.

One has to be careful not to get carried away, but occasionally this makes it easier to see the analogy between the classical case and more general rings.

1.38 (Infinitely near singularities). Let the notation be as in (1.36). Consider a curve $0 \in C \subset S$ with equation $(f = 0)$. There is a largest power $(x, y)^m$ of the maximal ideal that contains f; this m is the *multiplicity* of C at $0 \in S$, cf. (1.20.4).

Thus f modulo $(x, y)^{m+1}$ can be identified with a degree m homogeneous polynomial $f_m(x, y)$, called the *leading term* of f.

What happens to C under the blow-up?

On the chart $B_0 S_{v \neq 0} = \operatorname{Spec} R[\frac{y}{x}]$, the pullback of $f \in R$ is again $f \in R[\frac{y}{x}] = R[y_1]$. The change is that $y = y_1 x$, and thus $(x, y)^m \subset x^m (y_1, 1)^m$. This means that we can write f as $f = x^m f^1$ for some $f^1 \in R[y_1]$.

Thus the pullback of C contains the exceptional curve E with multiplicity m (defined by $(x^m = 0)$ on the $v \neq 0$ chart), and the *birational transform* of C, denoted by C_1, is defined by $(f^1 = 0)$ on the $v \neq 0$ chart:

$$\pi^* C = (\operatorname{mult}_0 C) \cdot E + C_1. \tag{1.38.1}$$

DEFINITION 1.39. The singularities of C_1 lying above $0 \in C$ are called the *infinitely near singularities* in the first infinitesimal neighborhood of $0 \in C$.

Similarly, the singularities in the first infinitesimal neighborhood of $\pi^{-1}(0) \subset C_1$ are called the infinitely near singularities in the second infinitesimal neighborhood of $0 \in C$, and so on.

Thus starting with a singular curve $0 \in C \subset S$, we get a towering system of infinitely near singularities on various blow-ups of S.

It is very easy to convince yourself through examples that each blow-up improves the singularities—that is, the infinitely near singularities are "better" than the original one—but it is nonetheless hard to come up with a good general statement about what actually improves. We will see several ways of doing it, but for now let us just see some simple results and examples.

LEMMA 1.40. *Let the notation be as above. The intersection points $C_1 \cap E$ are the roots of $(f_m = 0) \subset \mathbb{P}^1 \cong E$. More precisely, when counted with multiplicities, $C_1 \cap E = (f_m = 0)$. Thus,*

(1) *the intersection number $(C_1 \cdot E)$ equals $\operatorname{mult}_0 C$, and*

(2) $\sum_{p \in C_1 \cap E} \operatorname{mult}_p C_1 \leq \operatorname{mult}_0 C$.

Proof. Write $f = f_m + r$, where $r \in (x, y)^{m+1}$ (cf. (1.37)).

In the $v \neq 0$ chart $f^1 = f_m(y_1, 1) + x r_1$; hence the intersection points of C_1 and E, that is, the solutions of $f_1 = x = 0$, are the points $f_m(y_1, 1) = 0$ on E, proving (1). Furthermore, by (1.20.4), the multiplicity of C_1 at $(y_1 = a, x = 0)$ is $\leq$ the multiplicity of $y_1 = a$ as a root of $f_m(y_1, 1) = 0$, which gives (2). $\square$

EXAMPLE 1.41. Consider the singularity $C := (x^a = y^b)$. If $a < b$, then on the first infinitesimal neighborhood we find $C_1 = (x_1^a = y_1^{b-a})$.

If $a < b - a$, then on the second infinitesimal neighborhood we get $C_2 = (x_1^a = y_1^{b-2a})$ and so on.

Eventually we reach a stage where $b - ka \leq a$. Here we reverse the role of x and y and continue as before. Thus we see the usual Euclidean algorithm for the pair a, b carried out on exponents.

At some point we get an equation $x_s^c = y_s^c$. After one more blow-up we get c different smooth points. The singularity is thus resolved and we stop.

Thus far the resolutions constructed were abstract curves. In some cases the existence was only local, and in some other cases we ended up with a smooth curve or Riemann surface without any specific embedding into $\mathbb{C}^n$ or $\mathbb{P}^n$. In the Riemann surface case it is not even obvious that $\bar{C}$ has any embeddings into some $\mathbb{C}^n$.

It is frequently very useful to have a resolution that takes into account the ambient affine or projective space containing the curve C.

We summarize the above examples into an algorithm. Later we prove several cases when the algorithm does work.

ALGORITHM 1.42 (Weak embedded resolution algorithm). Let S_0 be a smooth surface over a perfect field and $C_0 \subset S_0$ a curve.

If $C_i \subset S_i$ is already constructed then pick any singular point $p_i \in C_i$, let $S_{i+1} \to S_i$ be the blow-up of p_i, and let $C_{i+1} \subset S_{i+1}$ be the birational transform of C_i.

For now we only state the weak embedded resolution theorem. The next six sections contain seven versions of its proof.

THEOREM 1.43 (Weak embedded resolution). *Let S be a smooth surface over a perfect field k and $C \subset S$ a reduced curve. After finitely many steps the weak embedded resolution algorithm (1.42) stops with $S_m \to S$ such that $C_m \subset S_m$ is smooth.*

It is often convenient to have an embedded resolution where not only is the birational transform of C smooth, but all the exceptional curves of the blow-ups behave as nicely as possible. We cannot make everything disjoint, but we can achieve the next best situation, simple normal crossing.

DEFINITION 1.44. Let X be a smooth variety and $D \subset X$ a divisor. We say that D is a *simple normal crossing* divisor (abbreviated as *snc* divisor) if every irreducible component of D is smooth and all intersections are transverse.

That is, for every $p \in X$ we can choose local coordinates $x_1, \ldots, x_n$ and natural numbers $m_1, \ldots, m_n$ such that $D = (\prod_i x_i^{m_i} = 0)$ in a neighborhood of x.

REMARK 1.45. A frequently occurring variant of this concept is that of a *normal crossing* divisor. Here we assume that for every $p \in X$ there are local analytic or formal coordinates $x_1, \ldots, x_n$ and natural numbers $m_1, \ldots, m_n$ such that $D = (\prod_i x_i^{m_i} = 0)$ in a local analytic or formal neighborhood of p.

Thus the nodal curve $y^2 = x^3 + x^2$ is a normal crossing divisor in $\mathbb{C}^2$ but not a simple normal crossing divisor. Indeed, we can write

$$y^2 - x^3 - x^2 = (y - x\sqrt{1+x})(y + x\sqrt{1+x})$$

as a power series, but $y^2 - x^3 - x^2$ is irreducible as a polynomial.

Be warned that in the literature the distinction between normal crossing and simple normal crossing is not systematic. In most cases the difference between them is a small technical matter, but occasionally it can cause difficulties.

ALGORITHM 1.46 (Strong embedded resolution algorithm). Let S_0 be a smooth surface over a perfect field k and $C_0 \subset S_0$ a curve.

If $\pi_i : S_i \to S$ is already constructed, then pick any point $p_i \in \pi_i^{-1}(C_0)$ where $\pi_i^{-1}(C_0)$ is not a simple normal crossing divisor, let $\sigma_{i+1} : S_{i+1} \to S_i$ be the blow-up of p_i and let $\pi_{i+1} := \pi_i \circ \sigma_{i+1} : S_{i+1} \to S_0$ be the composite.

THEOREM 1.47 (Strong embedded resolution). *Let S be a smooth surface over a perfect field k and $C \subset S$ a curve. After finitely many steps the strong embedded resolution algorithm (1.46) stops with $\pi_m : S_m \to S$ such that $\pi_m^{-1}(C)$ is a simple normal crossing divisor.*

1.48 (Proof of (1.43) $\Rightarrow$ (1.47)). The algorithm (1.46) does not specify the order of the blow-ups, but let us be a little more systematic first.

Starting with $C \subset S$, use (1.43) to get $C_m \subset S_m$ such that C_m is smooth. Let $E_m \subset S_m$ be the exceptional divisor of $S_m \to S$. Then C_m is smooth and E_m is a simple normal crossing divisor, but $C_m + E_m$ need not be a simple normal crossing divisor.

Let us now apply (1.43) again to $C_m + E_m \subset S_m$. We get $S_m^* \to S_m$ with exceptional divisor F^* such that the birational transform $C_m^* + E_m^* \subset S_m^*$ is smooth.

Since every irreducible component of $C_m + E_m$ is smooth, we see that every irreducible component of $C_m^* + E_m^*$ has only simple normal crossings with F^*. As $C_m^* + E_m^*$ is smooth, this implies that $C_m^* + E_m^* + F^*$ is a simple normal crossing divisor, and thus we have achieved strong embedded resolution.

We still need to show that blowing up in any other order also gets to strong embedded resolution. Even more, we claim that we always end up with a surface $S_n \to S$ that is dominated by S_m^*. That is, there is factorization $S_m^* \to S_n \to S$.

Assume that we already know that S_m^* dominates some S_i. If we blow up $p_i \in S_i$ as the next step of the algorithm (1.46), then $\pi_i^{-1}(C)$ does not have simple normal crossing at p, and hence the point p_i is also blown up when we construct S_m^* but maybe at a later stage. For blowing up points on a surface it does not matter in which order we blow them up, so S_m^* also dominates S_{i+1}, and we are done by induction. $\square$

Note that in getting $S_m^* \to S$ we may have performed some unnecessary blow-ups as well. Indeed, if $C \subset S$ has some simple normal crossing points, then these were blown up in constructing $S_m \to S$ but there is no need to blow these up in the algorithm (1.46).

1.49 (Digression on blow-ups over imperfect fields). I heartily recommend avoiding blowing up over imperfect fields unless it is absolutely necessary. Here are two examples to show what can happen.

Let u and v be indeterminates over a field k of prime characteristic p. Consider the affine plane $\mathbb{A}^2$ over the field $k(u,v)$.

Let $P \in \mathbb{A}^2$ denote the closed point corresponding to $(0, \sqrt[p]{v})$. The ideal of P is $(x, y^p - v)$. Thus the blow-up is given in $\mathbb{A}^2_{x,y} \times \mathbb{P}^1_{s,t}$ by the equation $xs - (y^p - v)t = 0$. Over the algebraic closure we can introduce a new coordinate $y_1 = y - \sqrt[p]{v}$, and the equation becomes $xs - y_1^p t = 0$. The resulting hypersurface is singular at $x = s = y_1 = 0$.

Let $Q \in \mathbb{A}^2$ denote the closed point corresponding to $(\sqrt[p]{u}, \sqrt[p]{v})$. The ideal of Q is $(x^p - u, y^p - v)$, and the blow-up is given by the equation $(x^p - u)s - (y^p - v)t = 0$. Over the algebraic closure we can introduce new coordinates $x_1 = x - \sqrt[p]{u}, y_1 = y - \sqrt[p]{v}$ and the equation becomes $x_1^p s - y_1^p t = 0$. The corresponding hypersurface is singular along the line $x_1 = y_1 = 0$.

On the other hand, both of the blow-ups are nonsingular hypersurfaces; that is, their local rings are regular.

In the first case, the only question is at the point with maximal ideal $(x, y^p - v, \frac{s}{t})$. The equation of the blown-up surface is $x\frac{s}{t} = y^p - v$, so x and $\frac{s}{t}$ are local coordinates on $B_P\mathbb{A}^2$.

In the second case, look at any point along the exceptional curve, say with maximal ideal $(x^p - u, y^p - v, \frac{s}{t} - a)$. The equation of the blown-up surface is

$$(x^p - u)(\tfrac{s}{t} - a) + a(x^p - u) = y^p - v,$$

and thus $x^p - u$ and $\frac{s}{t} - a$ are local coordinates on $B_Q \mathbb{A}^2$.

1.6. Embedded resolution, I: Global methods

Another observation of M. Noether is that it is easier to measure the singularities of a global curve. In effect, he considered the arithmetic genus of the curve. (Working on $\mathbb{P}^2$, he used the closely related intersection number between the affine curve $f(x, y) = 0$ and its polar curve $\partial f / \partial y = 0$.)

In hindsight it is not difficult to localize the proof, as we see in Section 1.8, but we start with the global version. For anyone familiar with basic algebraic geometry, the global version is faster, and the proof is a very nice application of intersection theory on surfaces. The local proofs are, however, more elementary.

We use the basic properties of intersection numbers of curves on smooth projective surfaces and of the canonical class of a surface. We start by recalling the relevant facts.

1.50 (Intersection numbers on smooth surfaces). Here is a summary of the properties that we use. For proofs see [Sha94, IV.1 and IV.3.2].

Let S be a smooth projective surface over a field k and C, D divisors on S. Then one can define their *intersection number*, $(C \cdot D)$ which has the following properties. (In fact, it is defined by the properties (1.50.1–3).)

(1) $(C \cdot D) \in \mathbb{Z}$ is bilinear.
(2) If $C_1 \sim C_2$ are linearly equivalent, then $(C_1 \cdot D) = (C_2 \cdot D)$.
(3) If C, D are effective and $C \cap D$ is finite, then

$$(C \cdot D) = \sum_{P \in C \cap D} \dim_k \mathcal{O}_{P,S}/(f_P, g_P),$$

where f_P (resp., g_P) is a local equation for C (resp., D) at P.
(4) Let $h : S' \to S$ be a birational morphism. Then $(h^* C \cdot h^* D) = (C \cdot D)$.
(5) Let $h : S' \to S$ be a birational morphism and $E \subset S'$ an h-exceptional divisor. Then $(f^* C \cdot E) = 0$.
(6) Let $h : S' \to S$ be the blow-up of a smooth k-point and $E \subset S'$ the exceptional curve. Then $(E \cdot E) = -1$.

The number $(C \cdot D)_P = \dim_k \mathcal{O}_{P,S}/(f_P, g_P)$ is called the *local intersection number* of C and D at P. (The local intersection number is defined only if $P \in C \cap D$ is an isolated point.)

1.51 (Canonical divisor). Let X be a smooth variety of dimension n. Divisors of rational differential n-forms on X are linearly equivalent, and they form the *canonical class* denoted by K_X; see, for instance, [Sha94, Sec.III.6.3]. It is a long-standing tradition to pretend sometimes that K_X is a divisor, but it is only a linear equivalence class.

One checks easily that $K_{\mathbb{P}^n} = -(n+1) \cdot$ (hyperplane class).

Let $h : S' \to S$ be the blow-up of a k-point P of a smooth surface defined over k and $E \subset S'$ the exceptional curve. Choose local coordinates (x, y) at P. An affine chart of S' is given by $y_1 = y/x, x_1 = x$ and

$$h^*(dx \wedge dy) = dx_1 \wedge d(x_1 y_1) = x_1 \cdot dx_1 \wedge dy_1.$$

Thus we conclude that

$$K_{S'} = h^* K_S + E. \qquad (1.51.1)$$

THEOREM 1.52 (Weak embedded resolution, I). *Let S_0 be a smooth projective surface over a perfect field k and $C_0 \subset S_0$ a reduced projective curve. After finitely many steps, the weak embedded resolution algorithm (1.42) stops with a smooth curve $C_m \subset S_m$.*

Proof. We look at the intersection number $C \cdot (C + K_S)$ and prove two properties:

(1.52.1) $C_{i+1} \cdot (C_{i+1} + K_{S_{i+1}}) < C_i \cdot (C_i + K_{S_i})$, and

(1.52.2) $C_i \cdot (C_i + K_{S_i})$ is bounded from below.

These two imply the termination of the blow-up process, and they even give a bound on the number of necessary steps.

Another variant of this method uses $h^1(C, \mathcal{O}_C)$ instead of $C \cdot (C + K_S)$; cf. [Har77, V.3.8].

The proof of (1.52.1) is a straightforward local computation using (1.53).

The proof of (1.52.2) is again not difficult. We discuss two methods that do not need the knowledge of resolution of curves.

Method 1 using duality theory. If you know enough duality theory (say, as in [Har77, Sec.III.8]) then you know that $\mathcal{O}_S(C + K_S)|_C$ is isomorphic to the dualizing sheaf ω_C. Then we claim that

$$\deg \omega_C \geq (-2) \cdot \#\{\text{irreducible geometric components of } C\}.$$

(With a little more care one could sharpen this to the number of connected geometric components.) This is easy if we know that C has a resolution, but here is an argument that does not rely on this.

If C is irreducible over an algebraically closed field k, pick two smooth points $p, q \in C$. From

$$0 \to \mathcal{O}_C(-p-q) \to \mathcal{O}_C \to k_p + k_q \to 0$$

we conclude that $h^1(C, \mathcal{O}_C(-p-q)) > 0$. Since $H^1(C, \mathcal{O}_C(-p-q))$ is dual to $\mathrm{Hom}(\mathcal{O}_C(-p-q), \omega_C)$, we conclude that $\deg \omega_C \geq \deg \mathcal{O}_C(-p-q) = -2$. (It is here that we use that C is reduced, and hence ω_C has no nilpotents. Otherwise, the homomorphism $\mathcal{O}_C(-p-q) \to \omega_C$ could not be used to bound the degree.)

If $C = \cup C_i$ are the irreducible geometric components, then

$$\big(C \cdot (C + K_S)\big) = \sum_i \big(C_i \cdot (C_i + K_S)\big) + \sum_{i \neq j} (C_i \cdot C_j) \geq \sum_i \big(C_i \cdot (C_i + K_S)\big)$$

and we are done. □

The next method is entirely elementary but gives a weaker bound.

Method 2 using differential forms. We prove two assertions.

(1.52.3) There is an injection $\Omega_C/(\text{torsion}) \hookrightarrow \mathcal{O}_S(C + K_S)|_C$.

(1.52.4) Let $f : C \to \mathbb{P}^1$ be a separable morphism of degree n. Then there is an injection $f^*\mathcal{O}_{\mathbb{P}^1}(-2) \cong f^*\Omega_{\mathbb{P}^1} \hookrightarrow \Omega_C$.

Assuming these, take a separable morphism $f_0 : C_0 \to \mathbb{P}^1$, say, of degree n. Since $C_i \to C_0$ are birational, we get degree n separable morphisms $f_i : C_i \to \mathbb{P}^1$ for every i. Thus $\deg \Omega_{C_i}/(\text{torsion}) \geq -2n$, and so $C_i \cdot (C_i + K_{S_i}) \geq -2n$, proving (1.52.2).

The proof of (1.52.4) is easy since differential forms can be pulled back and the pullback map is injective for a separable map.

Let us study the map in (1.52.3) in local coordinates x, y. Here Ω_S is generated by dx, dy and Ω_C is generated by the restrictions $dx|_C, dy|_C$.

What about $\mathcal{O}_S(C + K_S)|_C$? The local generator is $f^{-1}dx \wedge dy$, and then we take its residue along C. That is, we write

$$\frac{1}{f} dx \wedge dy = \frac{df}{f} \wedge \sigma,$$

and then $\sigma|_C$ is the local generator of $\mathcal{O}_S(C+K_S)|_C$. Thus we see that along the smooth points of C we can identify $\mathcal{O}_S(C + K_S)|_C$ with Ω_C, and even near singular points, $dx|_C, dy|_C$ give rational sections of $\mathcal{O}_S(C + K_S)|_C$. We only need to prove that they do not have poles. (In fact we see that they have zeros of quite high order.)

Since $df = (\partial f/\partial x)dx + (\partial f/\partial y)dy$, we get that

$$\frac{1}{f} dx \wedge dy = \frac{df}{f} \wedge \frac{dy}{\partial f/\partial x} = -\frac{df}{f} \wedge \frac{dx}{\partial f/\partial y},$$

and hence the local generator of $\mathcal{O}_S(C + K_S)|_C$ is

$$\sigma|_C = \frac{dy}{\partial f/\partial x} = -\frac{dx}{\partial f/\partial y}.$$

Therefore, $dx|_C = -(\partial f/\partial y)\sigma_C$ and $dy|_C = (\partial f/\partial x)\sigma_C$ both have zeros. □

LEMMA 1.53. *Let S be a smooth surface over a perfect field k and $C \subset S$ a projective curve. Let $p \in C$ be a closed point of degree d; that is, p is a conjugation-invariant set of d points in $C(\bar{k})$. Let $m = \text{mult}_p C$ be its multiplicity. Let $\pi : S' \to S$ denote the blow-up of p and $C' \subset S'$ the birational transform of C. Then*

(1) $(C' \cdot C') = (C \cdot C) - dm^2$,
(2) $(C' \cdot K_{S'}) = (C \cdot K_S) + dm$, and
(3) $(C' \cdot (C' + K_{S'})) = (C \cdot (C + K_S)) - dm(m - 1)$.

Proof. Over $\bar{k}$, the blow-up of p is just the blow-up of d distinct points, so it is enough to compute what happens under one such blow-up. Thus assume that k is algebraically closed.

We have already computed in (1.38.1) that $\pi^*C = C' + mE$, where $E \subset B_p S$ is the exceptional curve. Thus by (1.50.4–6) we get that

$$
\begin{aligned}
(C' \cdot C') &= (\pi^*C - mE) \cdot (\pi^*C - mE) \\
&= (\pi^*C \cdot \pi^*C) - 2m(\pi^*C \cdot E) + m^2(E^2) \\
&= (C \cdot C) - m^2.
\end{aligned}
$$

Similarly, using (1.51.1) we get that

$$
\begin{aligned}
(C' \cdot K_{S'}) &= \big((\pi^*C - mE) \cdot (\pi^*K_S + E)\big) \\
&= (C \cdot K_S) - m(E \cdot E).
\end{aligned}
$$

Finally (3) is a combination of (1) and (2). $\square$

1.7. Birational transforms of plane curves

If we start the embedded resolution process of the previous section with a plane curve $C \subset \mathbb{P}^2$, the method produces a smooth curve that sits in a plane blown up in many points. Classical geometers were foremost interested in plane curves, and from their point of view the natural problem was to start with a projective plane curve $f_1(x_0 : x_1 : x_2) = 0$ and to perform birational transformations of $\mathbb{P}^2$ that "improve" the equation step-by-step until one gets another equation $f_m(x_0 : x_1 : x_2) = 0$, which is the resolution of the original curve.

This is, however, too much to hope for, since many curves are not isomorphic to smooth plane curves at all (for instance, because a smooth plane curve of degree d has genus $(d - 1)(d - 2)/2$). Therefore, we have to settle for a model $f_m(x_0 : x_1 : x_2) = 0$, which has "relatively simple" singularities.

From our modern point of view, the method and its difficulties are the following.

Starting with $C \subset \mathbb{P}^2$, as a first resolution step in (1.52) we get $C_1 \subset B_p \mathbb{P}^2$. Usually C_1 is not even isomorphic to any plane curve, but we want to force it back into the plane $\mathbb{P}^2$. There are two complications.

- In our effort to put C_1 back into $\mathbb{P}^2$, we have to introduce new singular points.
- There is no "canonical" way of creating a plane curve out of C_1, and we have to make some additional choices. This makes the

process somewhat arbitrary, and it can happen that over finite fields there are no suitable choices at all.

With these limitations, there are two very nice classical solutions.

The one by M. Noether uses birational transformations $\mathbb{P}^2 \dashrightarrow \mathbb{P}^2$ to create a curve $C_m \subset \mathbb{P}^2$, which is birational to C and has only ordinary multiple points (1.54), that is, any number of smooth branches intersecting pairwise transversally.

The method of Bertini uses degree 2 maps $\mathbb{P}^2 \dashrightarrow \mathbb{P}^2$ to create a curve $C_m \subset \mathbb{P}^2$, which is birational to C and has only ordinary double points.

The Bertini method is lovely geometry, but now it survives only as a curiosity. The Noether method, however, still lies at the heart of our understanding of birational transformations between algebraic varieties.

DEFINITION 1.54. A singular point of a plane curve $p \in C$ is an *ordinary multiple point* if C has smooth branches intersecting pairwise transversally at p. Equivalently, in local coordinates (x, y) the equation of C is written as $\prod_i (a_i x + b_i y) +$ (higher terms) $= 0$, where none of the $a_i x + b_i y$ is a constant multiple of another.

DEFINITION 1.55 (Cremona transformation). Consider $\mathbb{P}^2$ with coordinates $(x : y : z)$. The standard *Cremona transformation* or *quadratic transformation* of $\mathbb{P}^2$ is the birational involution

$$\sigma : \mathbb{P}^2 \dashrightarrow \mathbb{P}^2 \quad \text{given by} \quad (x : y : z) \mapsto (x^{-1} : y^{-1} : z^{-1}) = (yz : xz : xy).$$

Thus σ is not defined at the three coordinate vertices $(0 : 0 : 1), (0 : 1 : 0), (1 : 0 : 0)$, and the three coordinate lines are contracted to points.

Alternatively, σ can be viewed as first blowing up the three coordinate vertices to get $B_3\mathbb{P}^2$, and then contracting the birational transforms of the three coordinate lines.

Consider a curve $C = (f(x : y : z) = 0)$, which has multiplicity a at $(0 : 0 : 1)$, b at $(0 : 1 : 0)$ and c at $(1 : 0 : 0)$. Then the birational transform of $C = (f(x : y : z) = 0)$ is

$$\sigma_* C = \left(x^{-c} y^{-b} z^{-a} f(yz : xz : xy) = 0 \right). \qquad (1.55.1)$$

More generally, let $p, q, r \in \mathbb{P}^2$ be three noncollinear points. We can choose a coordinate system such that p, q, r are the three coordinate vertices. Thus there is a quadratic transformation $\sigma_{p,q,r}$ with base points p, q, r. We blow up the points p, q, r, and then contract the birational transforms of the three lines ℓ_p, ℓ_q and ℓ_r through any two of p, q, r.

1.56 (The singularities of $\sigma_* C$). Let $C \subset \mathbb{P}^2$ be a plane curve of degree n and p, q, r noncollinear points with multiplicities a, b, c on C. (We allow a, b or c to be zero.) What are the singularities of the resulting new curve $(\sigma_{p,q,r})_* C$?

First, we perform the blow-up of the three points to get $B_{p,q,r}\mathbb{P}^2$. Here the singularities outside $\{p, q, r\}$ are unchanged, and the singularities at each of these 3 points are replaced by the singularities in their first infinitesimal neighborhood.

Then we contract the birational transforms of the three lines ℓ_p, ℓ_q and ℓ_r. This creates new singularities.

Assume now that the line ℓ_p is not contained in the tangent cones of C at q and r, and has only transverse intersections with C at other points. Then the birational transforms of C and of ℓ_p intersect transversally, and so $(\sigma_{p,q,r})_* C$ has an ordinary $(n - b - c)$-fold point at the image of ℓ_p.

ALGORITHM 1.57 (Birational transformation algorithm). Let k be an algebraically closed field and $C_0 \subset \mathbb{P}^2$ an irreducible plane curve. If $C_i \subset \mathbb{P}^2$ is already defined, pick any nonordinary point $p \in C$ and choose $q, r \in \mathbb{P}^2$ in a general position.

Let $\sigma_{p,q,r} : \mathbb{P}^2 \dashrightarrow \mathbb{P}^2$ be the quadratic transformation with base points p, q, r, and set $C_{i+1} := (\sigma_{p,q,r})_* C_i$.

A slight modification may be needed in positive characteristic. In order to avoid some inseparable projections, before each of the above steps we may have to perform an auxiliary quadratic transformation, where p is either a general point of $\mathbb{P}^2$ or a general point of C and $q, r \in \mathbb{P}^2$ are in general position.

The algorithm first appears in [Noe71] but without any proof. The first substantial proof is in [Nöt75]. (The first paper spells his name as M. Noether, the second as M. Nöther. Most of his papers use the first variant.)

THEOREM 1.58 (M. Noether, 1871). *Let k be an algebraically closed field and $C \subset \mathbb{P}^2$ an irreducible plane curve. Then the algorithm (1.57) eventually stops with a curve $C_m \subset \mathbb{P}^2$, which has only ordinary multiple points (1.54).*

Thus the composite of all the quadratic transformations of the algorithm is a birational map $\Phi : \mathbb{P}^2 \dashrightarrow \mathbb{P}^2$ such that $\Phi_*(C) \subset \mathbb{P}^2$ has only ordinary multiple points.

We give two proofs. The first one assumes that we already know embedded resolution as in (1.52). The second, following Noether's original approach, gives another proof of embedded resolution.

Proof using resolution. Pick any point $p \in C$, and let $\pi : C \to \mathbb{P}^1$ be the projection from p. In characteristic zero or if the characteristic does not divide $\deg C - \mathrm{mult}_p C$, the projection π is separable. Thus we can take two general lines through p such that they are not contained in the tangent cone of C at p and they have only transverse intersections with C at other points. Take also a third line (not through p) that intersects C transversally everywhere.

The genericity conditions of (1.56) on the corresponding quadratic transformation are satisfied. Thus we get a curve C_1 such that

- the singularities of C outside p are unchanged,
- the singularity at p is replaced by the singularities in its first infinitesimal neighborhood, and
- we have created one new ordinary multiple point of multiplicity $\deg C$ and two new ordinary multiple points of multiplicity $\deg C - \operatorname{mult}_p C$.

Thus we can follow along the resolution algorithm (1.42) and end up with a curve $C_m \subset \mathbb{P}^2$ with only ordinary multiple points.

In positive characteristic, we first perform an auxiliary quadratic transformation, where p is either a general point of $\mathbb{P}^2$ or a general point of C and $q, r \in \mathbb{P}^2$ are in general position. In the first case, we get a curve of degree $2 \deg C$, in the second case a curve of degree $2 \deg C - 1$. Either $2 \deg C - \operatorname{mult}_p C$ or $2 \deg C - 1 - \operatorname{mult}_p C$ is not divisible by the characteristic, and then we can proceed as above.   □

Noether's proof. We start as above, but we prove that eventually we get a curve with ordinary multiple points without assuming the existence of resolution.

We use a first approximation of the genus of the curve, which is called the *apparent genus* or *deficiency* in the classical literature. If $C \subset \mathbb{P}^2$ is a curve of degree d with singular points of multiplicity m_i, then this number is

$$g_{\mathrm{app}}(C) := \binom{d-1}{2} - \sum_i \binom{m_i}{2}.$$

Starting with C as above, let $d = \deg C$, $m_0 = \operatorname{mult}_p C$ and $m_i : i > 0$ be the multiplicities of the other multiple points of C. Using (1.55.1) and the above analysis of the singular points of C_1, we obtain that

- $\deg C_1 = 2d - m_0$,
- the other singular points of C give singular points of C_1 with the same multiplicity m_i,
- we get three new ordinary singularities with multiplicities $d, d - m_0, d - m_0$, and
- if $p \in C$ is not ordinary, then there is at least one new singular point corresponding to p.

If we ignore the points of the last type in the computation of the deficiency of C_1, then we get that

$$\begin{aligned} g_{\mathrm{app}}(C_1) \;&<\; \binom{2d-m_0-1}{2} - \sum_{i>0} \binom{m_i}{2} - \binom{d}{2} - 2\binom{d-m_0}{2} \\ &=\; \binom{d-1}{2} - \sum_i \binom{m_i}{2} = g_{\mathrm{app}}(C), \end{aligned}$$

where the middle equality is by explicit computation. The sequence of quadratic transformations thus stops if we show that $g_{\mathrm{app}}(C) \geq 0$ for any irreducible plane curve C.

Look at the linear system $|A|$ of all curves of degree $d - 1$ that have multiplicity $m_i - 1$ at every singular point $p_i \in C$. If $C = (f(x, y, z) = 0)$, then the curve $(\partial f / \partial x = 0)$ is in $|A|$, so $|A|$ is not empty. For any $A \in |A|$, the intersection $A \cap C$ consists of the points p_i, each with multiplicity $\geq m_i(m_i - 1)$ and a residual set R. These form a linear system $|R|$ on C. By Bézout's theorem (cf. [Sha94, IV.2.1]), the degree of $|R|$ is

$$\deg |R| = d(d - 1) - \sum_i m_i(m_i - 1) = 2g_{\mathrm{app}}(C) + 2(d - 1).$$

On the other hand,

$$\dim |R| = \dim |A| = \binom{d + 1}{2} - 1 - \sum_i \binom{m_i}{2} = g_{\mathrm{app}}(C) + 2(d - 1).$$

Since $\dim |R| \leq \deg |R|$ by (1.20.7), we conclude that $0 \leq g_{\mathrm{app}}(C)$. □

REMARK 1.59. Following the proof of (1.58) easily leads to Noether's genus formula for plane curves:

$$g(C) = \binom{d - 1}{2} - \sum_i \binom{m_i}{2},$$

where the sum runs through all infinitely near singular points of C.

ALGORITHM 1.60 (Bertini algorithm). Let k be an algebraically closed field and $C_0 \subset \mathbb{P}^2_k$ an irreducible plane curve. If $C_i \subset \mathbb{P}^2_k$ is already defined, pick any point $p_1 \in C$, which is neither smooth nor an ordinary node, and five other points $p_2, \ldots, p_6 \in \mathbb{P}^2$ in general position.

We use some basic properties of cubic surfaces; see [Rei88, §7] or [Sha94, Sec.IV.2.5].

The blow-up of $\mathbb{P}^2$ at these six points is a cubic surface S. Pick a general point $s \in S$, and let $\pi_s : S \dashrightarrow \mathbb{P}^2$ be the projection from s. Let

$$\sigma : \mathbb{P}^2 \dashrightarrow S \xrightarrow{\pi_s} \mathbb{P}^2$$

be the composite of the inverse of the blow-up followed by projection from s. Note that $\sigma : \mathbb{P}^2 \dashrightarrow \mathbb{P}^2$ has degree 2. In coordinates, σ is given by three general cubics through the six points $p_1, \ldots, p_6$.

Set $C_{i+1} := \sigma_* C_i$.

The following result is proved in [Ber94], but an added remark of F. Klein says that it was already known to Clebsch in 1869.

THEOREM 1.61 (Clebsch, 1869; Bertini, 1894). *Let k be an algebraically closed field and $C \subset \mathbb{P}^2_k$ an irreducible plane curve. Then the algorithm*

(1.60) eventually stops with a curve $C_m \subset \mathbb{P}^2$, which has only ordinary nodes.

Proof. As in the proof of (1.58) we again follow the resolution algorithm (1.52), and we only need to check the following lemma.

LEMMA 1.62. *Let $C \subset S \subset \mathbb{P}^3$ be a reduced curve on a smooth cubic surface and $\pi : S \dashrightarrow \mathbb{P}^2$ the projection from a general point $s \in S$. Then $\pi_* C \subset \mathbb{P}^2$ has the same singularities as C, plus a few ordinary nodes.*

Proof. We have to find a point $s \in S$ such that

(1) s is not on the tangent plane of C at any singular point of C,

(2) s is not on the tangent line of C at any smooth point of C,

(3) s is not on any secant line of C connecting a singular point of C with another point,

(4) s is not on any secant line of C connecting two smooth points of C with coplanar tangent lines, and

(5) s is not on a trisecant line of C.

The last condition would mean that the trisecant line of C is also a quadrisecant of S and hence one of the twenty-seven lines on S. For the first four cases we prove that the points in $\mathbb{P}^3$ where they fail is a subset of a union of linearly ruled surfaces. Since S is not ruled by lines, a general point on S satisfies all five conditions.

For (1) we have to avoid finitely many planes, for (2) the union of all tangent lines and for (3) the cones over C with vertex a singular point of C.

Condition (4) needs a little extra work, and I do it only in characteristic zero. The general case is left to the reader. For any smooth point $c \in C$, let L_c be the tangent line. Projecting C from L_c is separable, and hence there are only finitely many other points $c'_j \in C$ such that the tangent line at c'_j is coplanar with L_c. Thus s has to be outside the union of all lines $\langle c, c'_j \rangle$ as c runs through all smooth points of C.   □

If $C \subset \mathbb{P}^2$ is defined over a field k, in general the singular points are not defined over k, and so the steps of the algorithm (1.57) are not defined over k. A suitable modification works over any infinite perfect field. The proof uses some basic results about ruled surfaces.

THEOREM 1.63. *Let k be an infinite perfect field and $C \subset \mathbb{P}^2_k$ an irreducible plane curve. Then there is a birational map $\Phi : \mathbb{P}^2_k \dashrightarrow \mathbb{P}^2_k$ such that $\Phi_*(C) \subset \mathbb{P}^2_k$ has only ordinary multiple points.*

Proof. Pick a point $p \in \mathbb{P}^2 \setminus C$ that does not lie on any of the singular tangent cones of C and such that projection from p is separable.

As a first step, we blow up p and then we blow up all the nonordinary singular points of C. Then we contract the birational transforms of all lines passing through p and any of the nonordinary singular points of C. If there are m nonordinary singular points, we end up with a minimal ruled surface $\mathbb{F}_{m+1}$ and $C' \subset \mathbb{F}_{m+1}$ such that all the nonordinary singular points of C have been replaced by the singularities in their first infinitesimal neighborhoods.

If $E \subset \mathbb{F}_{m+1}$ denotes the negative section and $F \subset \mathbb{F}_{m+1}$ the fiber, then $|E + (m+1)F|$ is very ample outside E, and hence a general member $D \in |E + (m+1)F|$ intersects C' transversally.

Pick $m + 2$ general fibers $F_1, \ldots, F_{m+2}$ that intersect C' transversally, and blow up the intersection points $F_i \cap D$.

The birational transforms of F_i and of D are now all -1-curves, thus they can be contracted, and we get $\mathbb{P}^2$ after contraction.

Thus we get $C_1 \subset \mathbb{P}^2$ such that all the nonordinary singular points of C have been replaced by the singularities in their first infinitesimal neighborhoods, and we have a number of new ordinary singular points. Moreover, everything we did is defined over the ground field k.

Iterating this procedure eventually gives a curve with only ordinary singular points.   □

EXAMPLE 1.64. Let $C \subset \mathbb{P}^2_{\mathbb{F}_q}$ be an irreducible curve with ordinary multiple points. There are only $q + 1$ different tangent directions defined over $\mathbb{F}_q$ at any point of $\mathbb{P}^2$, and hence we conclude that the normalization of C has at most $(q+1)(q^2 + q + 1)$ points in $\mathbb{F}_q$.

On the other hand, every irreducible curve C over $\mathbb{F}_q$ is birational to a plane curve. Indeed, its function field $\mathbb{F}_q(C)$ is separable over some $\mathbb{F}_q(x)$ (cf. [vdW91, 19.7]), and so it can be given by two generators $\mathbb{F}_q(C) \cong \mathbb{F}_q(x, y)$. This provides the birational map of C to the affine plane over $\mathbb{F}_q$. There are curves over $\mathbb{F}_q$ with an arbitrary large number of $\mathbb{F}_q$-points, and so (1.63) fails for finite fields.

The following example, due to B. Poonen, gives very nice explicit curves with this property.

Claim 1.64.1. Let C_m be any curve over $\mathbb{F}_q$ of the form

$$\prod_{\deg f \leq m-1} (y - xf(x))(\text{unit at } (0,0)) + x^n(\text{unit at } (0,0)) = 0.$$

Then the normalization of C_m has q^m points over $(0,0)$ if $n > q^m + q^{m-1} + \cdots + 1$.

Indeed, let us see what happens under one blow-up. In the chart $y_1 = y/x, x_1 = x$ we get

$$\prod_{\deg f \leq m-1} (y_1 - f(x_1))(\text{unit at } (0,0)) + x_1^{n-q^m}(\text{unit at } (0,0)).$$

The terms where $f(0) \neq 0$ can be absorbed into the unit, so we have q^{m-1} points sitting over $(x_1, y_1) = (0, 0)$ and the same happens over any other point $(x_1, y_1) = (0, c)$.

Claim 1.64.2. The polynomial

$$\prod_{\deg f \leq m-1} \left(y - x f(x)\right) + x^n \left(y^{q^{m-1}} + x - 1\right) \quad \text{is irreducible.}$$

We use the Eisenstein criterion [vdW91, 5.5] over $\bar{\mathbb{F}}_q(x)[y]$ and the prime $x - 1$. Setting $x = 1$ we get

$$\prod_{\deg f \leq m-1} \left(y - f(1)\right) + y^{q^{m-1}} = \left(y^q - y\right)^{q^{m-1}} + y^{q^{m-1}} = y^{q^m},$$

so all but the leading coefficient (in y) is divisible by $x - 1$. The constant term (in y) is $x^n(x - 1)$, which is not divisible by $(x - 1)^2$.

Thus

$$C_m := \left(\prod_{\deg f \leq m-1} \left(y - x f(x)\right) + x^n \left(y^{q^{m-1}} - x + 1\right) = 0 \right) \subset \mathbb{A}^2$$

is an irreducible curve defined over $\mathbb{F}_q$ whose normalization has at least q^m points in $\mathbb{F}_q$.

1.65 (Aside on birational transforms of $(\mathbb{P}^2, C)$). In both of the above methods we have considerable freedom to choose the birational transformations, and a given curve C has many models with only ordinary multiple points. It is natural to ask if the method can be sharpened to get a nodal curve using birational maps of $\mathbb{P}^2$.

In the rest of the section we show that this is usually impossible. More precisely, we prove that if $C \subset \mathbb{P}^2$ is a curve of degree at least 7 such that every point has multiplicity $< \frac{1}{3} \deg C$ and $\phi : \mathbb{P}^2 \dashrightarrow \mathbb{P}^2$ is any birational map, then $\phi_* C$ is not nodal, save when C itself has only nodes and ϕ is an isomorphism.

The methods of the proof have more to do with the birational geometry of surfaces, so one may prefer to come back to this part after reading Section 2.2. See [KSC04, Chap.V] for further applications of this method.

Theorem 1.65.1. Let $C_1, C_2 \subset \mathbb{P}^2$ be two plane curves, and assume that $\text{mult}_p C_i < \frac{1}{3} \deg C_i$ for every $p \in C_i$.

Then every birational map $\Phi : \mathbb{P}^2 \dashrightarrow \mathbb{P}^2$ such that $\Phi_* C_1 = C_2$ is an isomorphism.

Proof. Assuming that $(\mathbb{P}^2, C_1)$ is birational to $(\mathbb{P}^2, C_2)$, by resolution for surfaces there is a common resolution

$$(S, C)$$

$$q_1 \swarrow \qquad\qquad \searrow q_2$$

$$(\mathbb{P}^2, C_1) \qquad\qquad (\mathbb{P}^2, C_2),$$

where $C = (q_i)_*^{-1} C_i$. Set $d_i = \deg C_i$, and assume that $d_1 \geq d_2$. Consider the linear system $|d_1 d_2 K_S + 3d_1 C|$.

We can first view this as

$$|d_1 d_2 K_S + 3d_1 C| = d_1 |d_2 K_S + 3(q_2)_*^{-1} C_2|$$

and use (1.65.2) to conclude that it has a unique effective member whose support is exactly the exceptional curves of q_2.

On the other hand, we can also view this as

$$|d_1 d_2 K_S + 3d_1 C| \supset d_2 |d_1 K_S + 3(q_1)_*^{-1} C_1| + 3(d_1 - d_2)C,$$

and thus again by (1.65.2) it has a member that contains C if $d_1 > d_2$, a contradiction. If $d_1 = d_2$, then $|d_1 d_2 K_S + 3d_1 C|$ has a unique effective member whose support is exactly the exceptional curves of q_1.

Thus q_1 and q_2 have the same exceptional curves, and so S is the blow-up of the graph of an isomorphism. $\qquad\square$

Lemma 1.65.2. Let $C \subset \mathbb{P}^2$ be a plane curve such that $\mathrm{mult}_p C < \frac{1}{3} \deg C$ for every $p \in C$. Let $q : S \to \mathbb{P}^2$ be any birational morphism, S smooth. Then the linear system $m|(\deg C)K_S + 3q_*^{-1}C|$ has a unique effective member for $m \geq 1$. Its support is exactly the exceptional curves of q.

Proof. For any curve C, the intersection number of $(\deg C)K_S + 3q_*^{-1}C$ with a general line in $\mathbb{P}^2$ is zero, and thus every effective member of $m|(\deg C)K_S + 3q_*^{-1}C|$ is supported on the exceptional curves of q. Exceptional 1-cycles cannot move in a linear system, so the linear system $m|(\deg C)K_S + 3q_*^{-1}C|$ is at most zero-dimensional.

The effectiveness of $m|(\deg C)K_S + 3q_*^{-1}C|$ follows from basic general results on singularities of pairs (see, for instance, [KSC04, Chap. VI]) applied to $(S, \frac{3}{\deg C}C)$. Here is a short proof.

Lemma 1.65.3. Let S be a smooth surface and Δ an effective $\mathbb{Q}$-divisor such that $\mathrm{mult}_p \Delta \leq 1$ for every $p \in S$. Let $f : S' \to S$ be a proper birational morphism, S' smooth, and let $\Delta' \subset S'$ denote the birational transform of Δ. Then the divisor $K_{S'} + \Delta' - f^*(K_S + \Delta)$ is effective. Furthermore, if $\mathrm{mult}_p \Delta < 1$ for every $p \in S$, then its support is the whole exceptional divisor.

Proof. By induction (and using (2.13)) it is enough to prove this for one blow-up. Let $\pi : B_p S \to S$ be the blow-up of the point $p \in S$ with exceptional curve E. Then

$$K_{B_p S} = \pi^* K_S + E \quad \text{and} \quad \pi^* \Delta = \Delta' + (\mathrm{mult}_p \Delta) \cdot E.$$

Thus

$$K_{S'} + \Delta' = \pi^*(K_S + \Delta) + (1 - \mathrm{mult}_p \Delta)E.$$

To continue with the induction we also need to show that

$$\mathrm{mult}_q\big(\Delta' + (1 - \mathrm{mult}_p\,\Delta)E\big) \le 1 \quad \text{for every } q \in B_pS.$$

This is obvious for $q \notin E$. If $q \in E$, then by (1.40)

$$
\begin{aligned}
\mathrm{mult}_q\big(\Delta' + (1 - \mathrm{mult}_p\,\Delta)E\big) \;&\le\; (E \cdot \Delta') + (1 - \mathrm{mult}_p\,\Delta) \\
&=\; \mathrm{mult}_p\,\Delta + (1 - \mathrm{mult}_p\,\Delta) = 1. \quad \square
\end{aligned}
$$

Example 1.65.4. Let $C_1 \subset \mathbb{P}^2$ be a general degree 6 curve with three nodes p, q, r. The Cremona transformation with base points p, q, r creates another degree 6 curve with three nodes.

The methods of the proof of (1.65.1) have been developed much further, and they lead to a general understanding of birational maps between varieties that are rational or are close to being rational. See [KSC04] for an elementary introduction to these techniques and results.

1.8. Embedded resolution, II: Local methods

Here we study the resolution of embedded curve singularities by direct local computations. We explicitly compute the chain of blow-ups and deduce a normal form for the singularity in case the multiplicity has not dropped in a sequence of blow-ups.

With each successive blow-up, these normal forms become more and more special, but they all require a suitable coordinate change. The final analysis is then to show that eventually such a coordinate change is impossible. The last step is actually quite delicate.

One could say that this approach relies on brute force, rather than a nice general idea. It is, however, extensions of this approach that have proved most successful in dealing with higher-dimensional resolution problems, so a careful study of it is very worthwhile.

1.66 (Idea of proof). Assume for simplicity that $S = \mathbb{A}_k^2$, the affine plane over a field k, and C is given by an equation $f(x, y) = 0$. We can write f as a sum of homogeneous terms

$$f(x, y) = f_m(x, y) + f_{m+1}(x, y) + \cdots.$$

If the lowest term f_m is not an mth power of a linear form, then by (1.40) the multiplicity drops after a single blow-up. Thus we can complete the proof by induction on the multiplicity.

The hard case is when f_m is an mth power, $f_m(x, y) = c(y - a_1 x)^m$. If k is a perfect field, then $a_1 \in k$ and by a linear change of coordinates we may assume that $f_m = y^m$. We call such an f to be in *normal form*.

(If the field k is not perfect and m is a power of the characteristic, it may happen that a_1 is not in k but in some inseparable extension of it. We will not deal with this case.)

To get the blow-up, set $y = y_1 x_1, x = x_1$. The birational transform of C is given by the equation

$$\left(f^1(x_1, y_1) = 0\right) \quad \text{where } f^1(x_1, y_1) := x_1^{-m} f(x_1, y_1 x_1).$$

At the next blow-up we have a multiplicity drop unless f^1 also has multiplicity m and its degree m part is the mth power of a linear form. We are now faced with a linear change of coordinates

$$(x_1, y_1) \mapsto (x_1, y_1 - a_2 x_1).$$

It turns out that if we make a quadratic coordinate change

$$(x, y) \mapsto (x, y - a_1 x - a_2 x^2),$$

then f and f^1 are both in normal form.

More generally, if the multiplicity does not drop for k blow-ups in a row, then there is a degree k coordinate change

$$(x, y) \mapsto (x, y - a_1 x - a_2 x^2 - \cdots - a_k x^k)$$

such that the inductively defined $f, f^1, \ldots, f^{k-1}$ are all in normal form.

If the multiplicity stays m for infinitely many blow-ups, then we would need a coordinate change

$$(x, y) \mapsto (x, y - a_1 x - a_2 x^2 - \cdots).$$

This is not a polynomial change, and hence we have to work now in the power series ring $\mathbb{C}[[x, y]]$.

It is easy to see that if $f, f^1, f^2, \ldots$ are all in normal form then

$$f = (y - a_1 x - a_2 x^2 - \cdots)^m (\text{unit}),$$

this of course only in $\mathbb{C}[[x, y]]$.

We are thus left with the question, is it possible that f is divisible by an mth power in $\mathbb{C}[[x, y]]$ but not divisible by an mth power in $\mathbb{C}[x, y]$?

This is easily settled using the differential criterion of smoothness (1.72), but it is precisely this question that causes the most trouble when we try to replace a surface over k with the spectrum of a 2-dimensional regular local ring; see (1.103).

THEOREM 1.67. *Let S_0 be a smooth surface over a perfect field k and $C_0 \subset S_0$ a reduced curve. Then the embedded resolution algorithm (1.42) terminates with $C_m \subset S_m$, where C_m is smooth*

Proof. The question is local, so we may assume that $p_0 \in C_0$ is the only singular point.

As a first step, we use induction on the multiplicity. Let $\mathrm{mult}_{p_0} C_0 = m$. As we saw in (1.40), $\sum_{p \in C_1 \cap E} \mathrm{mult}_p C_1 \leq \mathrm{mult}_0 C$, where summation is over all infinitely near singularities in the first infinitesimal neighborhood of $p_0 \in C_0$. Thus we can use induction, unless there is only one singularity after blow-up and it also has multiplicity m.

We repeat the procedure. If the multiplicity drops after some blow-ups, then we can use induction. Thus the proof of (1.67) is complete if we can prove the following.

LEMMA 1.68. *Let the notation be as above. There is no infinite sequence of blow-ups*

$$(p_0 \in C_0 \subset S_0) \leftarrow (p_1 \in C_1 \subset S_1) \leftarrow (p_2 \in C_2 \subset S_2) \leftarrow \cdots$$

where

(1) $\mathrm{mult}_{p_i} C_i = m \geq 2$ *for every* i,
(2) C_{i+1} *is obtained from* C_i *by blowing up* p_i, *and*
(3) p_{i+1} *is the unique singular point of* C_{i+1} *lying above* p_i.

A sequence as in (1.68)

$$(p_0 \in C_0 \subset S_0) \leftarrow \cdots \leftarrow (p_k \in C_k \subset S_k)$$

is called a *length k blow-up sequence* of multiplicity m.

Let us now pass to the algebraic side. Let R_0 be the local ring of S_0 at p_0 and $f^0 \in R_0$ a local equation of C_0.

As we noted in (1.40), if the multiplicity does not drop then one can choose local parameters $x, y \in R_0$ such that the leading term of f^0 is y^m. That is, $f^0 \in (y^m) + (x, y)^{m+1}$.

The following is the key computation.

PROPOSITION 1.69. *With the above notation, there is a length k blow-up sequence of multiplicity m iff there are local parameters (x, y_k) such that*

$$f^0 \in (x^{k+1}, y_k)^m.$$

Moreover, if the above length k blow-up sequence extends to a length $k + 1$ blow-up sequence of multiplicity m, then we can choose $y_{k+1} = y_k - a_k x^k$ for some $a_k \in R_0$ such that $f^0 \in (x^{k+2}, y_{k+1})^m$.

Proof. If $f^0 \in (x^{k+1}, y_k)^m$, then after k blow-ups C_k lies on the affine chart $\mathrm{Spec}\, R_k$, where $R_k = R_0[\frac{y_k}{x^k}]$. Indeed,

$$(x^{k+1}, y_k)^m = x^{km} \left(x, \frac{y_k}{x^k} \right)^m;$$

thus we can write $f^0 = x^{mk} f^k$, and the unique singular point of $C_k = (f^k = 0)$ is at the maximal ideal $(x, \frac{y_k}{x^k})$. What happens at the next blow-up?

Here we get the ring $R_{k+1} = R_0[\frac{y_k}{x^{k+1}}]$ and by assumption C_{k+1} has multiplicity m at some point of the exceptional curve, say, at the maximal ideal $(x, \frac{y_k}{x^{k+1}} - a_{k+1})$. Thus $f^{k+1} \in x^m (x, \frac{y_k}{x^{k+1}} - a_{k+1})^m$, which gives

$$f^0 \in x^{m(k+1)} \left(x, \frac{y_k}{x^{k+1}} - a_{k+1} \right)^m = (x^{k+2}, y_k - a_{k+1} x^{k+1})^m.$$

This is exactly what we wanted, but sloppy notation helped us here. Indeed, we need that $f^0 \in (x^{k+2}, y_k - a_{k+1}x^{k+1})^m R_0$, but we proved that

$$f^0 \in (x^{k+2}, y_k - a_{k+1}x^{k+1})^m R_0 \left[\frac{y_k}{x^{k+1}}\right].$$

Fortunately, these two are equivalent by (1.70).   □

LEMMA 1.70. *Let R be a unique factorization domain and $x, y \in R$ distinct primes. Then*

$$R \cap (x^b, y)R\left[\frac{y}{x^a}\right] = (x^b, y).$$

Proof. Every element of $(x^b, y)R[\frac{y}{x^a}]$ can be written as

$$s = \frac{y^c}{x^{ac}}r_1 x^b + \frac{y^d}{x^{ad}}r_2 y,$$

where $x^a \nmid r_1$ if $c \geq 1$ and $x^a \nmid r_2$ if $d \geq 1$.

Assume that $s \in R$. If $c = 0$, then $\frac{y^d}{x^{ad}}r_2 y \in R$, which happens only for $d = 0$, and then $s \in (x^b, y)$.

If $c \geq 1$, then we can write $s = ys_1$, where

$$s_1 = \frac{y^{c-1}}{x^{ac}}r_1 x^b + \frac{y^d}{x^{ad}}r_2 \in R\left[\frac{1}{x}\right].$$

Multiplying by y cannot cancel out a denominator that is a power of x; thus $s_1 \in R$, and so $s \in (y) \subset (x^b, y)$.   □

These coordinate changes can all be put together if we pass from the ring R to its *completion*, denoted by $\hat{R}$. See [AM69, Chap.10] for the definition and its basic properties.

COROLLARY 1.71. *Let S_0 be a smooth surface over a perfect field k and R_0 its local ring at a point p_0 with completion $\hat{R}_0$. There is an infinite blow-up sequence of multiplicity m iff there are local parameters x and $y_\infty = y - \sum_{k \geq 1} a_k x^k \in \hat{R}_0$ such that*

$$f^0 \in (y_\infty)^m.$$

Proof. Applying (1.69) we get that $f^0 \in (x^{k+1}, y_\infty)^m$ for every k. Thus, working in $\hat{R}_0/(y_\infty)^m$, we obtain that $f^0 \in \cap_k (x^{k+1})$, and the latter intersection is zero by Krull's intersection theorem (cf. [AM69, 10.17]).   □

1.72 (Going back from $\hat{R}_0$ to R_0). By assumption S_0 is a smooth surface and C_0 a reduced curve; thus it has an isolated singularity at p_0. That is, p_0 is an isolated solution of $f^0 = \partial f^0/\partial x = \partial f^0/\partial y = 0$. Thus

$$\hat{R}/\left(f^0, \tfrac{\partial f^0}{\partial x}, \tfrac{\partial f^0}{\partial y}\right) = R/\left(f^0, \tfrac{\partial f^0}{\partial x}, \tfrac{\partial f^0}{\partial y}\right)$$

is finite dimensional. On the other hand, if $f^0 = y_\infty^m$(unit), then

$$\left(f^0, \frac{\partial f^0}{\partial x}, \frac{\partial f^0}{\partial y}\right) \subset (y_\infty^{m-1}),$$

a contradiction.   □

It is straightforward to see that the method of the proof gives the following stronger result.

THEOREM 1.73. *Let S be a 2-dimensional regular scheme over a field and $C \subset S$ a reduced curve. Assume that C has finitely many singular points $c_i \in C$ and the residue fields $k(c_i)$ are perfect. Then*

(1) *either the weak embedded resolution algorithm terminates with $C_m \subset S_m$, where C_m is smooth,*

(2) *or the completion $\widehat{\mathcal{O}}_{c_i,C}$ is nonreduced for some i.*

We will use this result when $S = \operatorname{Spec} R$ is the spectrum of a 2-dimensional complete regular local ring with perfect residue field. Then $\widehat{\mathcal{O}}_{c_i,C} = \mathcal{O}_{c_i,C}$, so the second case never happens, and we always get a strong embedded resolution.

1.9. Principalization of ideal sheaves

So far we have dealt with a single curve $C \subset S$, but now we consider linear systems of curves on S. Equivalently, we are considering a finite-dimensional vector space of sections $V \subset H^0(S, L)$ of a line bundle L on S. Let $L_V \subset L$ be the subsheaf that they generate. Then $I_V := L_V \otimes L^{-1}$ is an ideal sheaf, and the main result says that after some blow-ups the pullback of any ideal sheaf becomes locally principal.

THEOREM 1.74 (Principalization of ideal sheaves). *Let S be a smooth surface over a perfect field k and $I \subset \mathcal{O}_S$ an ideal sheaf. Then there is a sequence of point blow-ups $\pi : S_m \to S$ and a simple normal crossing divisor $F_m \subset S_m$ such that*

$$\pi^* I = \mathcal{O}_{S_m}(-F_m),$$

where $\pi^* I \subset \mathcal{O}_{S_m}$ denotes the ideal sheaf generated by the pullback of local sections of I. (This is denoted by $\pi^* I \cdot \mathcal{O}_{S_m}$ or by $I \cdot \mathcal{O}_{S_m}$ in [Har77, p.163].)

Applied to $V \subset H^0(S, L)$ and $I = I_V$ as above, we get that V, as a subspace of $H^0(S, \pi^* L) = H^0(S, L)$, generates the subsheaf $\pi^* L(-F_m)$, which itself is locally free. Thus V, as a subspace of $H^0(S, \pi^* L(-F_m))$, gives a base point–free linear system, and we obtain the following.

COROLLARY 1.75 (Elimination of base points). *Let $|D|$ be a finite-dimensional linear system of curves on a smooth surface S defined over a* *perfect field k. Then there is a sequence of point blow-ups $\pi : S_m \to S$ such that*

$$\pi^*|D| = F_m + |M|,$$

where $F_m \subset S_m$ is a simple normal crossing divisor and $|M|$ is a base point–free linear system. $\square$

Let $f : S \dashrightarrow \mathbb{P}^n$ be a rational map and $|H|$ the linear system of hyperplanes. The base points of $f^*|H|$ are precisely the points where f is not a morphism. Thus using (1.75) for $f^*|H|$ we conclude the following.

COROLLARY 1.76 (Elimination of indeterminacies). *Let S be a smooth surface defined over a perfect field k and $f : S \dashrightarrow \mathbb{P}^n$ a rational map. Then there is a sequence of point blow-ups $\pi : S_m \to S$ such that the composite*

$$f \circ \pi : S_m \to S \dashrightarrow \mathbb{P}^n \quad \text{is a morphism.} \quad \square$$

1.77 (Proof of (1.74)). For a projective surface over an infinite field, a simple proof is given in [Sha94, IV.3.3]. Here we give a local argument in the spirit of Section 1.8.

There are only finitely many points where I is not of the form $\mathcal{O}_S(-D)$ for some simple normal crossing divisor D. The problem is local at these points, so we may assume that S is affine. Write $I = (g_1, \ldots, g_t)$ and set $D_i = (g_i = 0)$.

First we apply the strong embedded resolution theorem (1.47) to $D_1 + \cdots + D_t \subset S$. (To avoid too many subscripts, we keep denoting the blown-up surface again by S.) Thus we are reduced to the case where $D_1 + \cdots + D_t \subset S$ is a simple normal crossing divisor.

At a smooth point of $D_1 + \cdots + D_t$ with local equation $x = 0$, the pullbacks of the g_i are locally given as $x^{m_i}(\text{unit})$, so the pullback ideal is locally principal with generator x^m, where $m = \min\{m_i\}$.

We now have to look more carefully at the finitely many nodes of $\text{red}(D_1 + \cdots + D_t)$.

If $s \in S$ is such a node, we can choose local coordinates x, y and natural numbers A_i, C_i such that $x^{A_i} y^{C_i} = 0$ is a local equation for D_i. I is locally principal at s iff there is a j such that

$$A_j \leq A_i \quad \forall i \quad \text{and} \quad C_j \leq C_i \quad \forall i. \tag{1.77.1}$$

After one blow-up, we get two new nodes, and the new local equations are

$$\left(\tfrac{x}{y}\right)^{A_i} y^{A_i + C_i} = 0, \quad \text{resp.,} \quad x^{A_i + C_i} \left(\tfrac{y}{x}\right)^{C_i} = 0.$$

We need to prove that in any sequence of blow-ups we eventually get a locally principal ideal. The condition (1.77.1) shows that this is a purely combinatorial question, which is settled by (1.79) when we use $k_s = 0$ for every s. $\square$

REMARK 1.78. This proof of (1.74) works for any 2-dimensional regular scheme S, where strong embedded resolution holds. In particular, by (1.73), it holds when $S = \operatorname{Spec} R$ is the spectrum of a 2-dimensional complete regular local ring with perfect residue field.

Hironaka proposed some combinatorial games that aim to describe the above process where player one chooses a subvariety to blow up and player two decides which chart of the blow-up to consider next [Hir72]. In the simplest case we need, we have only the origin to blow up, so player one has no choices at all, and the game is not very exciting. (The numbers k_s are added with later applications in mind (2.68).)

LEMMA 1.79 (2-dimensional Hironaka game). *Let k_s be a sequence of rational numbers and (A_i, C_i) finitely many pairs of rational numbers. We define inductively a sequence $(A_i(s), C_i(s))$, where $A_i(0) = A_i, C_i(0) = C_i$ and*

$$\bigl(A_i(s+1), C_i(s+1)\bigr) = \begin{cases} either & \bigl(A_i(s) + C_i(s) + k_s, C_i(s)\bigr) & \forall i, \\ or & \bigl(A_i(s), A_i(s) + C_i(s) + k_s\bigr) & \forall i. \end{cases}$$

Then, for $s \gg 1$, there is a j such that for very i, $A_j(s) \leq A_i(s)$ and $C_j(s) \leq C_i(s)$.

Proof. The simplification under blow-ups is more transparent if we set $a(s) := \min_i\{A_i(s)\}$, $c(s) := \min_i\{C_i(s)\}$ and write $a_i(s) = A_i(s) - a(s)$, $c_i(s) = C_i(s) - c(s)$. If we set $m(s) = \min\{a_i(s) + c_i(s)\}$, then the transformation rules are

$$\bigl(a_i(s+1), c_i(s+1)\bigr) = \begin{cases} either & \bigl(a_i(s) + c_i(s) - m_s, c_i(s)\bigr) & \forall i, \\ or & \bigl(a_i(s), a_i(s) + c_i(s) - m_s\bigr) & \forall i. \end{cases}$$

We see that $\min_i\{a_i(s+1) + c_i(s+1)\} \leq \min_i\{a_i(s) + c_i(s)\}$, and strict inequality holds unless the minimal value m_s is achieved only for a pair $(0, m_s)$ if we use the first rule and for a pair $(m_s, 0)$ if we use the second rule. By our definitions, there is also a pair of the form $(a_k(s), 0)$ or $(0, c_k(s))$. For these the above rules give

$$\begin{aligned} \bigl(a_k(s+1), c_k(s+1)\bigr) &= \bigl(a_k(s) - m_s, 0\bigr) \quad \text{or} \\ \bigl(a_k(s+1), c_k(s+1)\bigr) &= \bigl(0, c_k(s) - m_s\bigr). \end{aligned}$$

Thus we see that for $\min_i\{a_i(s) + c_i(s)\} > 0$ the pair

$$\Bigl(\min_i\{a_i(s) + c_i(s)\}, \min_{i:c_i(s)=0}\{a_i(s)\} + \min_{i:a_i(s)=0}\{c_i(s)\}\Bigr)$$

decreases lexicographically at each step. As we have positive rational numbers with bounded denominators, we get that $(a_j(s), c_j(s)) = (0, 0)$ for some j for $s \gg 1$. Thus $A_j(s) \leq A_i(s)$ and $C_j(s) \leq C_i(s)$ for every i. $\square$

1.10. Embedded resolution, III: Maximal contact

Let S be a smooth surface and $C \subset S$ a reduced curve, singular at a point p. As we saw at the beginning of the proof of (1.67), in order to resolve the singularities of C, it is enough to reduce $\text{mult}_p C$ by repeated blow-ups. The situation was completely analyzed in (1.67). The main difficulty was that we had to adjust our coordinate system at $p \in S$ repeatedly to put the equation into normal form for more and more blow-ups. At the end we needed a formal or analytic coordinate change to take care of all possible blow-ups.

The method of maximal contact allows us, at least in characteristic zero, to write down an "optimal" coordinate system at the beginning. Even better, we write down a smooth curve $p \in H \subset S$, which plays the same role as the formal curve $(y_\infty = 0)$ played in the proof of (1.67), and we do not need to worry about coordinate systems any longer.

The method of maximal contact plays a crucial role in the proof of the higher-dimensional resolution theorems in Chapter 3.

DEFINITION 1.80. Let S be a smooth surface, $C \subset S$ an irreducible curve and $p \in C$ a point of multiplicity m. A smooth curve $p \in H \subset S$ is called a *curve of maximal contact* of C at p if the following holds.

Let S' be a smooth surface and $\pi : S' \to S$ a birational morphism. Denote by C' (resp., H') the birational transform of C (resp., H) on S'. Then H' contains every point $p' \in \pi^{-1}(p)$ such that $\text{mult}_{p'} C' \geq m$.

The following two theorems show how to use maximal contact curves for resolution.

THEOREM 1.81. *Let S be a smooth surface, $C \subset S$ an irreducible curve and $p \in C$ a point of multiplicity m. Assume that C has a curve of maximal contact $p \in H \subset S$, and let $M := (C \cdot H)_p$ be the local intersection number.*

Then after at most M/m blow-ups the multiplicity of the birational transform of C is less than m at every preimage of p.

THEOREM 1.82. *Let S be a smooth surface over a field of characteristic zero and (x, y) local coordinates at a point $p \in S$. Let $C = (f(x, y) = 0) \subset S$ be an irreducible curve of multiplicity m, and assume that y^m appears in $f(x, y)$ with nonzero coefficient. Then*

$$H := \left(\frac{\partial^{m-1} f}{\partial y^{m-1}} = 0 \right) \quad \text{is a curve of maximal contact for } C \text{ at } p.$$

Proof of (1.81). Let $(p_0 \in C_0 \subset S_0) \leftarrow \cdots \leftarrow (p_k \in C_k \subset S_k)$ be a length k blow-up sequence of multiplicity m as in (1.68). Let $H_i \subset S_i$ denote the birational transform of H. By assumption $p_i \in H_i$ for every $i \leq k$. By (1.53.1),

$$m \leq (C_{i+1} \cdot H_{i+1})_{p_{i+1}} \leq (C_i \cdot H_i)_{p_i} - m \quad \text{for every } 0 \leq i \leq k-1,$$

which implies that $k \leq M/m$. $\square$

    Proof of (1.82). On the blow-up $\pi : B_p S \to S$ consider the coordinate chart given by $x_1 = x, y_1 = y/x$. The equation of the birational transform C_1 of C is

$$f^1(x_1, y_1) := x_1^{-m} f(x_1, x_1 y_1).$$

By the chain rule, this gives that

$$\frac{\partial f^1}{\partial y_1} = x_1^{-m} \cdot \frac{\partial f}{\partial y} \cdot \frac{\partial y}{\partial y_1} = x_1^{-(m-1)} \cdot \frac{\partial f}{\partial y}(x_1, x_1 y_1).$$

That is, the y-derivative of the birational transform is the birational transform of the y-derivative. Repeating this $(m-1)$-times, we get that

$$\frac{\partial^{m-1} f^1}{\partial y_1^{m-1}} = x_1^{-1} \frac{\partial^{m-1} f}{\partial y^{m-1}}(x_1, x_1 y_1).$$

This shows that the birational transform of H is

$$H_1 = \left(\frac{\partial^{m-1} f^1}{\partial y_1^{m-1}} = 0\right).$$

If f^1 has multiplicity m at some point $p_1 \in B_p S$, then its $(m-1)$st partial derivatives are all zero at p_1. Thus H_1 passes through every preimage of p, where C_1 has multiplicity m. (We know by (1.40.2) that there is at most one such point.)

    If C_1 again has a point of multiplicity m, then $f^1(0, y_1) = c(y_1 - \alpha)^m$. Thus after a coordinate translation $(x_1, y_1) \mapsto (x_1, y_1 - \alpha)$ (which commutes with differentiations) the equation of C^1 is $(f^1 = 0)$, and y_1^m appears in $f^1(x_1, y_1)$ with nonzero coefficient.

    Hence H_1 is defined by the equation $\partial^{m-1} f^1 / \partial y_1^{m-1} = 0$ and we can repeat the argument for further blow-ups.

    Thus H is a curve of maximal contact for C. $\square$

REMARK 1.83. It should be stressed that the above curve of maximal contact H depends on the coordinate system (x, y). The partial derivatives of f are not coordinate invariants, but for every j the ideal $D^j(f)$ generated by all jth partial derivatives of all elements of the ideal (f) is well defined. The H in (1.82) is the zero set of a typical element of $D^{m-1}(f)$. Any other smooth curve defined by some $\phi \in D^{m-1}(f)$ also has maximal contact with C.

1.11. Hensel's lemma and the Weierstrass preparation theorem

    As we saw already in Newton's method, resolving curve singularities is very closely related to solving algebraic equations $f(x, y) = 0$. In general there are no power series solutions, only Puiseux series solutions.

In this section we deal with two results—Hensel's lemma and the Weierstrass preparation theorem—that show how far one can go in simplifying power series $F(x, y)$ if we do not want to pass to Puiseux series.

Let us start with the simplest case, which is also easy to derive from Newton's method.

LEMMA 1.84. *Let $F(x, y) \in K[[x]][y]$ be a polynomial in y whose coefficients are power series in x. Assume that α is a simple root of $F(0, y) = 0$. Then $F(x, y) = 0$ has a unique power series solution*

$$y = \sum_{i=0}^{\infty} b_i x^i \quad \text{with } b_0 = \alpha.$$

Proof. We try to find the coefficients of the assumed power series solution $y = b_0 + b_1 x + b_2 x^2 + \cdots$ inductively.

That is, first, we solve modulo x; this amounts to solving the ordinary polynomial equation

$$F(0, y) = 0 \quad \text{or, equivalently,} \quad F(x, y) \equiv 0 \mod (x).$$

By assumption, $b_0 = \alpha$ is a solution. Then we try to get a linear term $b_1 x$ such that

$$F(x, b_0 + b_1 x) \equiv 0 \mod (x^2).$$

Next comes the quadratic part $b_2 x^2$ and so on.

Assume that we already found $b_0 + b_1 x + \cdots + b_m x^m$ such that

$$F(x, b_0 + b_1 x + \cdots + b_m x^m) \equiv 0 \mod (x^{m+1}).$$

Next we want to get b_{m+1} such that

$$F(x, b_0 + b_1 x + \cdots + b_m x^m + b_{m+1} x^{m+1}) \equiv 0 \mod (x^{m+2}).$$

Set $y_m(x) := b_0 + b_1 + \cdots + b_m x^m$. By Taylor's theorem

$$F\big(x, y_m(x) + b_{m+1} x^{m+1}\big) \equiv F\big(x, y_m(x)\big) + \frac{\partial F}{\partial y}\big(x, y_m(x)\big) b_{m+1} x^{m+1},$$

modulo $(x^{2(m+1)})$. By assumption

$$F\big(x, y_m(x)\big) \equiv C_{m+1} x^{m+1} \mod (x^{m+2})$$

for some constant C_{m+1} and

$$\frac{\partial F}{\partial y}\big(x, y_m(x)\big) x^{m+1} \equiv \frac{\partial F}{\partial y}(0, b_0) x^{m+1} \mod (x^{m+2}).$$

Thus we need to solve the equation

$$C_{m+1} + \frac{\partial F}{\partial y}(0, b_0) b_{m+1} = 0.$$

This has a unique solution if $\partial F/\partial y(0, b_0) \neq 0$, that is, when b_0 is a simple root of $F(0, y) = 0$. Thus inductively we find a unique power series solution

$$y = \sum_{i=0}^{\infty} b_i x^i.$$

If, however, $\partial F/\partial y(0, b_0) = 0$, that is, when b_0 is a multiple root of $F(0, y) = 0$, then we have no solution in general. ☐

REMARK 1.85. The Newton iteration method for finding roots of a real or complex polynomial $f(x)$ is the rule

$$x_{i+1} = x_i - \frac{f(x_i)}{f'(x_i)}.$$

One should recognize the method of (1.84) as following this iteration and truncating at each step. We work in a ring $K[[x]]$ instead of a field, and hence the invertibility of $f'(x_i)$ is a bigger problem than usual.

If $F(0, y)$ has only multiple roots, we have no power series solutions in general, but we can separate the different roots of $F(0, y)$ from each other without passing to Puiseux series. This is the content of Hensel's lemma.

THEOREM 1.86 (Hensel's lemma). *Let $F(x, y) \in K[[x]][y]$ be a polynomial in y whose coefficients are power series in x. Assume that*

$$F(0, y) = g(y)h(y),$$

where g and h are relatively prime. Then we can write

$$F(x, y) = G(x, y)H(x, y), \quad where \ G(0, y) = g(y) \ and \ H(0, y) = h(y).$$

Moreover, there is a unique solution G, H such that

$$\deg_y\big(G(x, y) - g(y)\big) < \deg_y g(y).$$

Proof. As before, we try to find inductively

$$G_m(x, y) := \sum_{i=0}^{m} g_i(y)x^i \quad \text{and} \quad H_m(x, y) = \sum_{i=0}^{m} h_i(y)x^i$$

such that

$$G_m(x, y)H_m(x, y) \equiv F(x, y) \mod (x^{m+1}).$$

The starting case is $G_0(x, y) = g(y)$ and $H_0(x, y) = h(y)$. In order to go from m to $m + 1$, we need to solve, modulo (x^{m+2}), the equation

$$\big(G_m(x, y) + g_{m+1}(y)x^{m+1}\big)\big(H_m(x, y) + h_{m+1}(y)x^{m+1}\big) \equiv F(x, y).$$

By assumption,

$$G_m(x, y)H_m(x, y) \equiv F(x, y) + f_{m+1}(y)x^{m+1} \mod (x^{m+1})$$

for some $f_{m+1}(y)$, and thus we are down to solving

$$g(y)h_{m+1}(y) + h(y)g_{m+1}(y) = f_{m+1}(y). \qquad (1.86.1)$$

Since g and h are relatively prime, $g(y)a(y) + h(y)b(y) = 1$ is solvable, so we can take $h_{m+1}(y) = a(y)f_{m+1}(y)$ and $g_{m+1}(y) = b(y)f_{m+1}(y)$.

Moreover, if $u(y)g(y) + v(y)h(y) = f_{m+1}(y)$ is any solution, then we can uniquely write $v(y) = g(y)s(y) + g_{m+1}(y)$, where $\deg g_{m+1} < \deg g$. This gives a unique solution of (1.86.1) with $\deg g_{m+1} < \deg g$ for every $m \geq 1$. In this case (1.86.1) implies that $\deg_y H \leq \deg_y F - \deg_y g$; thus H is also a polynomial in y.

Thus there is a unique solution where $\deg_y(G(x,y) - g(y)) < \deg_y g(y)$. $\qquad\square$

It is worthwhile to look at the above proof to see if one can replace the polynomial ring $K[[x]][y]$ by some other ring. The first observation is that we used only that $K[[x]][y]/(x^m)$ is a polynomial ring in y, and so we could have worked in the ring $K[y][[x]]$ to start with. (Note that this is indeed a bigger ring. For instance, $\sum x^i y^i \in K[y][[x]] \cap K[x][[y]]$, but it is not in $K[[x]][y]$.) Without any change in the proof, we get the following general form.

PROPOSITION 1.87 (Abstract Hensel lemma). *Let R be a ring and $F(x) := \sum_{i=0}^{\infty} r_i x^i$ a power series in x with coefficients in R. Assume that*

$$r_0 = g_0 h_0 \quad and \quad g_0 R + h_0 R = R.$$

Then we can write

$$F(x) = G(x)H(x), \quad where \ G(0) = g_0 \ and \ H(0) = h_0.$$

Moreover, if we fix representatives r^* for every residue class $R/(g_0)$, then there is a unique solution where $G(x) = g_0 + \sum_{i\geq 1} g_i^* x^i$. $\qquad\square$

As a simple consequence, we obtain the Weierstrass preparation theorem. which shows that any power series $F(x,y)$ can be made into a polynomial in one of the variables.

COROLLARY 1.88 (Weierstrass preparation theorem). *Let $F(x,y) \in K[[x,y]] = K[[y]][[x]]$ be a power series. Assume that y^n appears in F with nonzero coefficient and n is the smallest such exponent. Then one can write $F(x,y)$ uniquely as*

$$F(x,y) = \left(y^n + g_{n-1}(x)y^{n-1} + \cdots + g_0(x)\right) \cdot (unit),$$

where $g_i \in K[[x]]$.

Proof. We apply (1.87) to $R = K[[y]]$. Here we can write $F(0,y) = y^n \cdot u(y)$, where $u(y)$ is a unit; set $g_0 := y^n$ and $h_0 := u$. The unit h_0 alone generates $K[[y]]$, and every $\phi(y) \in K[[y]]$ can be written uniquely as

$\phi(y) = y^n \cdot s(y) + u(y) \cdot t(y)$, where $\deg t < n$. Indeed, $t(y)$ is the sum of the first n terms of $\phi(y) \cdot u(y)^{-1}$. $\qquad\qquad\qquad\qquad\qquad\qquad\square$

1.89 (Embedded resolution using Weierstrass' theorem). Let $0 \in C$ be a singular point of a curve on a smooth surface S. If $\mathrm{mult}_0\, C = m$, then we can choose local coordinates such that y^m appears in the equation of C with nonzero coefficient. Thus by (1.88) in the completion $\hat{\mathcal{O}}_{0,S}$ we can write the equation as

$$\hat{C} = \left(y^m + g_{m-1}(x)y^{m-1} + \cdots + g_0(x) = 0\right).$$

If the characteristic does not divide m, then by a change of coordinates $y \mapsto y - \frac{1}{m}g_{m-1}$ we get the simpler equation

$$\hat{C} = \left(y^m + a_{m-2}(x)y^{m-2} + a_{m-3}(x)y^{m-3} + \cdots + a_0(x) = 0\right). \quad (1.89.1)$$

This means that $(y = 0)$ is a curve of maximal contact (1.80), and the computation of the blow-ups is especially easy in this case.

The assumption $\mathrm{mult}_0\, C = m$ means that $\mathrm{mult}_0\, a_{m-j} \geq j$. On the first blow-up everything happens on the chart with coordinates $y_1 = y/x, x_1 = x$. The birational transform of $\hat{C}$ is

$$\hat{C}_1 = \left(y_1^m + x_1^{-2}a_{m-2}(x_1)y_1^{m-2} + x_1^{-3}a_{m-3}(x_1)y_1^{m-3} + \cdots + x_1^{-m}a_0(x_1) = 0\right).$$

The form of the equation is unchanged, and $\mathrm{mult}_0\, \hat{C}_1 \geq m$ iff $\mathrm{mult}_0\, a_{m-j} \geq 2j$ for every j. Thus after k blow-ups we have coordinates $y_k = y/x^k, x_k = x$ and the birational transform $\hat{C}_k$ of $\hat{C}$ is given by the equation

$$\left(y_k^m + x_k^{-2k}a_{m-2}(x_k)y_k^{m-2} + x_k^{-3k}a_{m-3}(x_k)y_k^{m-3} + \cdots + x_k^{-mk}a_0(x_k) = 0\right).$$

From this we conclude that

$$\mathrm{mult}_0\, \hat{C}_k = m \quad \text{iff} \quad k \leq \min\left\{\frac{\mathrm{mult}_0\, a_{m-i}(x)}{i} : i = 2, \ldots, m\right\}.$$

Thus we can read off the precise number of blow-ups needed to lower the multiplicity from the equation (1.89.1).

This is the simplest illustration of the general theory of coefficient ideals, which is used in the proof of resolution in higher dimensions.

In order to make this approach into a proof of resolutions, one needs to analyze the relationship between blowing up the completion $\hat{\mathcal{O}}_{0,S}$ and blowing up S. Everything works out as expected, and the easy details are left to the reader.

ASIDE 1.90 (Weighted blow-ups). One can achieve the first k steps of the above blow-up sequence by just one *weighted blow-up*. In this case this amounts to blowing up the ideal (y, x^k). The resulting surface has a singular point, but the birational transform of the curve does not pass through it, so it does not interfere with the resolution of the curve.

Weighted blow-ups and more generally toric methods (see [Ful93]) work very efficiently if we know beforehand the optimal choice of the coordinate system or if the equation has only a few monomials. See [KSC04, Sec.6.5] for a nice example and for complete definitions.

The proof of the following form of Hensel's lemma is also left to the reader.

PROPOSITION 1.91. *For* $F(x,y) \in K[[x]][y]$*, let* $y_{\mathrm{app}}(x) \in K[[x]]$ *be an approximate solution; that is,*

$$F\big(x, y_{\mathrm{app}}(x)\big) \equiv 0 \mod (x^{m+1})$$

for some m*. Assume that*

$$m \geq 2n, \quad \text{where} \quad n := \mathrm{mult}_x \frac{\partial F}{\partial y}\big(x, y_{\mathrm{app}}(x)\big).$$

Then there is a solution $y(x)$ *such that*

$$y(x) \equiv y_{\mathrm{app}}(x) \mod (x^{m+1-n}). \quad \Box$$

It is straightforward to generalize the results of this section to the case where instead of one variable x we have several variables $x_1, \ldots, x_n$.

THEOREM 1.92 (General Hensel lemma). *Let* R *be a ring and*

$$F(x_1, \ldots, x_n) := \sum_{i_1, \ldots, i_n} r_{i_1, \ldots, i_n} x_1^{i_1} \cdots x_n^{i_n}$$

a power series in $x_1, \ldots, x_n$ *with coefficients in* R*. Assume that*

$$r_{0, \ldots, 0} = gh \quad \text{and} \quad gR + hR = R.$$

Then we can write

$$F(\mathbf{x}) = G(\mathbf{x})H(\mathbf{x}), \quad \text{where } G(0, \ldots, 0) = g \text{ and } H(0, \ldots, 0) = h.$$

Moreover, if we fix representatives r^* *for every residue class* $R/(g)$*, then there is a unique solution, where* $g_{i_1, \ldots, i_n} = g^*_{i_1, \ldots, i_n}$ *whenever* $(i_1, \ldots, i_n) \neq (0, \ldots, 0)$. $\quad \Box$

As before, applying this to $K[[y]][[x_1, \ldots, x_n]]$ we get the following.

THEOREM 1.93 (Weierstrass preparation theorem). *Let* K *be a field and*

$$F(x_1, \ldots, x_n, y) \in K[[x_1, \ldots, x_n, y]]$$

a power series. Assume that y^m *appears in* F *with nonzero coefficient and* m *is the smallest such. Then one can write* F *uniquely as*

$$F = \big(y^m + g_{m-1}(\mathbf{x})y^{m-1} + \cdots + g_0(\mathbf{x})\big) \cdot (\text{unit}),$$

where $g_i(\mathbf{x}) \in K[[x_1, \ldots, x_n]]$. $\quad \Box$

1.12. Extensions of $K((t))$ and algebroid curves

In this section we describe finite field extensions of the field of formal Laurent series $K((t))$ over an algebraically closed field K. Equivalently, we solve equations $f(t, y) = 0$, where $f(t, y) \in K((t))[y]$.

This is the setting of Newton's method of the rotating ruler, and it is straightforward to rewrite it in a more algebraic setting.

More generally, we can work with germs of formal curves; that is, we consider $F(x, y) = 0$, where $F(x, y) \in K[[x, y]]$ is a formal power series. These are also called *algebroid* curves or singularities.

THEOREM 1.94. *Let K be an algebraically closed field and $K((t))$ the field of formal Laurent series. Let $L \supset K((t))$ be a finite degree field extension. Assume that* char K *does not divide* $\deg |L/K((t))|$.

Then $L = K((t^{1/m}))$, where $m = \deg |L/K((t))|$.

Proof. First we establish that $L \subset K((t^{1/M}))$ for some M, using induction on $\deg |L/K((t))|$.

Pick any $y \in L \setminus K((t))$. It satisfies a minimal equation

$$\sum_{i=0}^{n} a_i(t) y^i = 0, \quad \text{where } a_i(t) \in K((t)) \text{ and char } K \nmid n.$$

We can multiply through by a large power of t to assume that $a_i(t) \in K[[t]]$. Then multiply the equation by $a_n(t)^{n-1}$, and substitute $y \mapsto a_n(t)^{-1} y$ to get a monic equation

$$y^n + b_{n-1}(t) y^{n-1} + \cdots + b_0(t) = 0, \quad \text{where } b_i(t) \in K[[t]].$$

Finally, since the characteristic does not divide n, we can substitute $y \mapsto y - \frac{1}{n} b_{n-1}(t)$ to get

$$F(t, y) := y^n + c_{n-2}(t) y^{n-2} + \cdots + c_0(t) = 0, \quad \text{where } c_i(t) \in K[[t]].$$

(This last transformation is a seemingly minor convenience, but the proof really hinges on it. We use it through the following elementary observation: A polynomial $y^n + a_{n-2} y^{n-2} + \cdots + a_0 \in K[y]$ is not an nth power of a linear polynomial if char $K \nmid n$, unless $a_{n-2} = \cdots = a_0$.)

Let us look now at the leading part in t, which is $F(0, y) \in K[y]$, and distinguish two cases.

Case 1. If $F(0, y)$ has at least two distinct roots, then it can be written as $F(0, y) = F_1(0, y) F_2(0, y)$, where F_1 and F_2 are relatively prime. Thus by the Hensel lemma (1.86), F has a corresponding nontrivial factorization $F(t, y) = F_1(t, y) F_2(t, y)$, contradicting the assumption that F is a minimal polynomial of y.

Case 2. If F has only one root, then $F(0, y) = (y - \alpha)^n$ for some $\alpha \in K$. Comparing the coefficients of y^{n-1} we see that $\alpha = 0$, and so $F(0, y) = y^n$. Thus t divides every $c_i(t)$.

If by chance $t^{r(n-i)} | c_i(t)$ for some $r > 0$ and every i, then we can substitute $y \mapsto t^r y$ and divide by t^{nr} to get the equation

$$\left(\frac{y}{t^r}\right)^n + \frac{c_{n-2}(t)}{t^{2r}}\left(\frac{y}{t^r}\right)^{n-2} + \cdots + \frac{c_0(t)}{t^{nr}} = 0, \quad \text{where} \quad \frac{c_i(t)}{t^{r(n-i)}} \in K[[t]].$$

The previous argument with the Hensel lemma now shows that $t | c_i(t)/t^{r(n-i)}$ for every i.

We do not get much out of this if we take r to be an integer, but allowing r to be rational turns out to be very useful.

The largest value of r allowed without ending up with poles in some $c_i(t)/t^{r(n-i)}$ is

$$r = \min_i \left\{ \frac{\text{mult}_0 \, c_i(t)}{n-i} \right\}.$$

The minimum is achieved for some $i_0 \leq n - 2$, and then $r = u/v$ with $v | n - i_0 \leq n - 2$. By setting $s = t^{1/v}$, $y_1 = y/t^r$ and $d_i(t^{1/v}) = c_i(t)/t^{r(n-i)}$, we get an equation

$$G(y_1, s) := y_1^n + d_{n-2}(s)y^{n-2} + \cdots + d_0(s) = 0.$$

As before, the $d_i(s)$ are power series in s, and $d_{i_0}(s)$ has a nonzero constant term.

This means that $G(y_1, 0)$ is different from y_1^n. Hence by the argument in Case 1, $G(y_1, s)$ factors nontrivially over $K((s)) = K((t^{1/v}))$. (It is not hard to see that char $K \nmid v$, but let us not worry about this right now.)

In any case, at least one of the factors has a degree that is not divisible by char K, so by induction on the degree, we conclude that $y \in K((t^{1/M}))$ for some M. More precisely, at first we know this only for some conjugate of y, but $K((t^{1/M}))/K((t))$ is a normal extension, so in fact we get $y \in K((t^{1/M}))$.

This implies that the degree of $L(t^{1/M})/K((t^{1/M}))$ is a proper divisor of the degree of $L/K((t))$. By induction we conclude that $L \subset K((t^{1/M}))$ (for a possibly larger M).

In characteristic zero, $K((t^{1/M}))/K((t))$ is a Galois extension with cyclic Galois group, and so every intermediate field is of the form $K((t^{1/m}))$ for some $m | M$. Hence, $L = K((t^{1/m}))$, where $m = \deg |L/K((t))|$.

In positive characteristic write $M = M'p^a$, where $p = \text{char } K$ and p does not divide M'. Then $K((t^{1/M'}))$ is the unique maximal separable subextension of $K((t^{1/M}))$. Since char K does not divide $\deg |L/K((t))|$, $L/K((t))$ is separable, and thus $L \subset K((t^{1/M'}))$. Now we can complete the argument as in the characteristic zero case. $\qquad\square$

By contrast, degree p extensions in characteristic p have no simple description. First of all, there is a unique inseparable extension of degree p, obtained by adjoining all pth roots.

In many problems, however, the main difficulty is caused by degree p separable extensions and *wild ramification*.

1.95 (Artin-Schreier theory). cf. [Ser79, X.3] Let k be a field of characteristic p and k'/k a degree p Galois extension. Then k' is the splitting field of a polynomial $y^p - y - a$, and $a \in k$ is unique modulo elements of the form $\{b^p - b : b \in k\}$.

Assume next that $k = K((t))$ and K is perfect. If $g \in tK[[t]]$, then

$$-g = \left(\sum_{i \geq 1} g^{p^i}\right)^p - \left(\sum_{i \geq 1} g^{p^i}\right),$$

but only the t^{-pc} terms can be eliminated from the negative exponents. Thus we conclude that a degree p Galois extension of $K((t))$ can be uniquely written as the splitting field of a polynomial

$$y^p - y - \sum a_i t^{-i}, \quad \text{where } a_i \in K \text{ and } a_i = 0 \text{ if } p|i.$$

This is in sharp contrast with the characteristic zero case, where there is a unique extension of a given degree if K is algebraically closed.

It is instructive to see what happens with $y^p - y - t^{-1}$ if we try to follow the method of the proof of (1.94). We have to rewrite it as a power series. This is achieved by multiplying by t^p and substituting $y \mapsto yt^{-1}$ to get

$$y^p - t^{p-1}y - t^{p-1} = 0.$$

At the first step we substitute $s = t^{1/p}$ to get $y^p - s^{p(p-1)}y - s^{p(p-1)}$. After $p - 1$ blow-ups we get the equation

$$\left(\frac{y}{s^{p-1}} - 1\right)^p - s^{p-1}\left(\frac{y}{s^{p-1}} - 1\right) - s^{p-1} = 0.$$

Thus we keep adjoining the roots t^{1/p^n}, but the equation stays irreducible all the time.

We can also restate our results in terms of Puiseux expansions.

THEOREM 1.96 (Puiseux expansion). *Let $F(x,y) \in K[[x,y]]$ be an irreducible formal power series with $F(0,0) = 0$. Assume that y^m appears in $F(x,y)$ with nonzero coefficient and m is the smallest such. Assume furthermore that $\operatorname{char} K$ does not divide m.*

Then $F(x,y) = 0$ has a solution of the form

$$y = \sum_{i=1}^{\infty} a_i x^{i/m} \in K[[x^{1/m}]],$$

called the Puiseux expansion of y.

In particular,

$$x \mapsto t^m, y \mapsto \sum_{i=1}^{\infty} a_i t^i \quad \text{give an injection} \quad K[[x,y]]/F(x,y) \hookrightarrow K[[t]],$$

which makes $K[[t]]$ into the normalization of $K[[x, y]]/F(x, y)$. Alternatively,

$$\operatorname{Spec}_K K[[t]] \to \operatorname{Spec}_K K[[x, y]]/(F) \quad \text{given by} \quad (t^m, \textstyle\sum_{i=1}^{\infty} a_i t^i)$$

is the resolution of singularities of the algebroid curve $F(x, y) = 0$.

Proof. By the Weierstrass preparation theorem (1.88), up to a unit, we can write F in the form

$$y^m + a_{m-1}(x)y^{m-1} + \cdots + a_0(x).$$

By assumption this is an irreducible, and hence $K((x))(y)/K((x))$ is a field extension of degree m. Therefore, by (1.94), $K((x))(y) \cong K((x^{1/m}))$. In other words, y can be written as a power series in $x^{1/m}$.

The normalization of $K[[x]]$ in $K((x^{1/m}))$ is clearly $K[[x^{1/m}]]$, which shows the second assertion. $\qquad\square$

1.13. Blowing up 1-dimensional rings

In this section we deal with the most general case of resolution in dimension 1. Can one resolve arbitrary 1-dimensional schemes?

Already in dimension 1 we run into the foundational mess of commutative ring theory. Starting with the works of E. Noether and Krull, for a few decades it seemed that Noetherian rings form the right abstract setting for affine algebraic geometry. Especially after the pathological examples of Nagata discovered in the 1950s, it became evident that many of the finer properties of the coordinate rings of algebraic varieties fail for general Noetherian rings. These discoveries culminated in a long series of proposed classes of rings, none of which seemed to have taken care of all problems completely.

At the end of the section we give three examples of 1-dimensional Noetherian integral domains, which show how bad the situation can get.

On the positive side, we deal with the resolution of 1-dimensional rings with only finitely many maximal ideals, called *semilocal rings*.

Even in this special case the answer is not always positive, but we show that any example where the blowing-up process goes on forever also has other bad properties.

DEFINITION 1.97. Let S be a 1-dimensional semi-local ring with maximal ideals $m_1, \ldots, m_k$. Set $I(S) := m_1 \cap \cdots \cap m_k$. The blow-up of all maximal ideals is the scheme

$$B_{I(S)} \operatorname{Spec} S := \operatorname{Proj}_S(S + I + I^2 + \cdots).$$

If I has n generators, then $B_{I(S)} \operatorname{Spec} S \subset \mathbb{P}^{n-1}_{\operatorname{Spec} S}$ is a closed subscheme. The fiber of $B_{I(S)} \operatorname{Spec} S \to \operatorname{Spec} S$ over the closed point $\operatorname{Spec}(S/m_i)$ is

$$\operatorname{Proj}_{\operatorname{Spec}(S/m_i)}(S/m_i + m_i/m_i^2 + \cdots),$$

which is zero-dimensional. Thus $B_{I(S)} \operatorname{Spec} S \to \operatorname{Spec} S$ is finite, and so $B_{I(S)} \operatorname{Spec} S$ is affine with coordinate ring denoted by BS. The ring BS is again semi-local and finite over S.

We do not have to worry about embedded points when we blow up.

LEMMA 1.98. *Let M be a zero-dimensional ideal. Then $B_{I(S)} \operatorname{Spec} S$ has no embedded points and $B_{I(S)} \operatorname{Spec} S \cong B_{I(S/M)} \operatorname{Spec}(S/M)$.*

Proof. By the Artin-Rees lemma (cf. [AM69, 10.9]), $I(S)^k \cap M = 0$ for $k \gg 1$. Thus only finitely many of the summands in $\sum_{j \geq 0} I(S)^j$ contain a nonzero element of M, and so the natural map

$$\sum_{i \geq 0} I(S)^i \to \sum_{i \geq 0} (I(S)/M)^i$$

is an isomorphism in all but finitely many degrees. Hence, they have the same projectivization.

Similarly, embedded points on $B_{I(S)} \operatorname{Spec} S$ would come from embedded primes in $I(S)^k$ for every k, and thus the blow-up has no embedded points supported on the preimage of $\operatorname{Supp} \mathcal{O}_{S/I}$.   □

LEMMA 1.99. *Let S be a 1-dimensional semi-local ring without zero-dimensional torsion. Then $BS \cong S$ iff S is regular.*

Proof. The blow-up $B_{I(S)} \operatorname{Spec} S \to \operatorname{Spec} S$ is an isomorphism on the complement of the closed points, and hence the kernel of $S \to BS$ consists of elements whose support is zero-dimensional. These do not effect the blow-up, and hence $S \to BS$ is injective iff $\operatorname{Spec} S$ has no embedded points.

The fiber of $B_{I(S)} \operatorname{Spec} S \to \operatorname{Spec} S$ over $\operatorname{Spec}(S/m_i)$ is

$$\operatorname{Proj}_{\operatorname{Spec}(S/m_i)} S/m_i + m_i/m_i^2 + \cdots ,$$

and thus the fiber is isomorphic to $\operatorname{Spec}(S/m_i)$ iff $m_i^j/m_i^{j+1} \cong S/m_i$ for $j \gg 1$; that is, iff m_i^j is a principal ideal for $j \gg 1$.

Assume that $m_i^j = (y_j)$ and $m_i^{j+1} = (y_{j+1})$ are both principal ideals. Thus $y_j m_i = (y_{j+1})$. Pick any $y_1 \in m_i$ such that $y_j y_1 = y_{j+1}$. If m_i is not principal, take any $x \in m_i \setminus (y_1)$. Then $y_j x = z y_{j+1}$ for some $z \in S$, and thus $y_j(x - z y_1) = 0$ but $x - z y_1 \neq 0$.

As y_j generates a power of a maximal ideal, it is invertible at every generic point, and thus $x - z y_1$ has zero-dimensional support, a contradiction.   □

ALGORITHM 1.100 (General blow-up algorithm). Let S_0 be a semi-local ring of dimension 1. If S_i is already defined, set $S_{i+1} := BS_i$.

The main theorem is the following description of those 1-dimensional semi local rings that have a resolution.

THEOREM 1.101. [Kru30, Satz.7] *Let S be a 1-dimensional, semi-local ring without embedded points. The following are equivalent.*

(1) *The blow-up sequence $S_0 = S, S_1, S_2, \ldots$ of (1.100) stabilizes after finitely many steps, and the rings $S_m = S_{m+1} = \cdots$ are regular.*

(2) *S is reduced, and the normalization $\bar{S}$ is finite over S.*

(3) *The completion $\hat{S}$ is reduced.*

Proof. We prove the equivalences in five steps.

- (1) for $S \Leftrightarrow$ (2) for S,
- (1) for $\hat{S} \Leftrightarrow$ (2) for $\hat{S}$,
- (1) for $S \Leftrightarrow$ (1) for $\hat{S}$,
- (2) for $S \Rightarrow$ (3), and
- (3) $\Rightarrow$ (2) for $\hat{S}$.

As we noted, each S_i is contained in $\bar{S}$ and by (1.99) $S_j \to S_{j+1}$ is an isomorphism iff S_j is regular. Thus (1) and (2) are equivalent both for S and $\hat{S}$, proving the first two steps.

By the Artin-Rees lemma (cf. [AM69, 10.9]) $\widehat{BS} = B\hat{S}$, and this implies that the blow-up sequence starting with $\hat{S}$ is the completion of the blow-up sequence starting with S. A semi-local ring is regular iff its completion is regular (cf. [AM69, 11.24]), and thus one of these sequences stabilizes after m blow-ups iff the other stabilizes after m blow-ups, proving step three.

To see then fourth step, assume that $\bar{S}$ is finite over S. By the Artin-Rees lemma, the $I(S)$-adic topology on S agrees with the topology induced by the $I(S)$-adic topology on $\bar{S}$. Thus $\hat{S}$ is a subring of the completion of $\bar{S}$.

The completion of a semilocal ring is the sum of the completions of the localizations. In our case $\bar{S}$ is regular by the equivalence of (2) and (1) and the completion of a regular local ring is again regular, and hence an integral domain (cf. [AM69, 11.24]) since they have the same m/m^2. Thus the completion of $\bar{S}$ has no torsion, and the same holds for its subring $\hat{S}$.

Finally, the last step is a special case of a general result of Nagata that for any complete local ring R its normalization $\bar{R}$ is finite over R. We sketch a proof of it below.   □

THEOREM 1.102. [Nag62, Sec.32] *Let S be a complete, local, Noetherian integral domain and $F \supset Q(S)$ a finite field extension of its quotient field. Then the normalization of S in F is finite over S.*

Outline of the proof. By and large we follow the method of (1.33).

First, this is easy if $F/Q(S)$ is separable and S is normal; see [AM69, 5.17]. Thus we can assume that $Q(S)$ has positive characteristic. A hard step is to show that S can be written as an S/m_S-algebra (cf. [Mat89, Sec.29]). Another easy result (cf. [AM69, Ch.11.Exrc.2]) shows that S

is finite over a complete regular local ring. This reduces us to studying the normalization of $k[[x_1, \ldots, x_n]]$ in finite purely inseparable extensions. Every such extension is contained in one of the form

$$k^{p^{-m}}[[x_1^{p^{-m}}, \ldots, x_n^{p^{-m}}]],$$

but this is finite over $k[[x_1, \ldots, x_n]]$ only if $k^{p^{-m}}/k$ is finite.

This covers completions of local rings of an algebraic variety but fails in general. See [Nag62, Sec.32] on how to fix this problem or [Mat70, Sec.31] for a different approach. □

EXAMPLE 1.103. [Nag62, App.Exmp.3] Here we give an example of a DVR S and a degree p field extension of its quotient field $Q(S)$ such that the normalization of S in F is not finite over S.

This also leads to an example of 1-dimensional, Noetherian local domain R such that $\bar{R}$ is not finite over R.

Let k be a field of characteristic p, and let $F = k(x_1, x_2, \ldots)$ be generated by infinitely many algebraically independent elements. Set $K = k(x_1^p, x_2^p, \ldots)$. Note that $F^p \subset K$ and F has infinite degree over K. Let S be the union of all the power series rings $E[[t]]$, where E runs through all subextensions $K \subset E \subset F$ that have finite degree over K.

S is a 1-dimensional regular local ring with maximal ideal (t) and residue field $S/(t) \cong F$.

Set $R := S[y]/(y^p - \sum_{i \geq 1} x_i^p t^{pi})$. Note that $\sum_i x_i^p t^{pi} \in S$ since $x_i^p \in K$ for every i, but its pth root $\sum_{i \geq 1} x_i t^i \notin S$ since its coefficients $x_1, x_2, \ldots$ generate the whole F. This implies that $y^p - \sum_i x_i^p t^{pi}$ is irreducible in $S[y]$, and thus R is an integral domain.

CLAIM 1.104. The normalization of R is not finite over R, and hence the normalization of S in the degree p extension $Q(R)/Q(S)$ is not finite over S.

Proof. The quotient field of S is $Q(S) := \cup_n t^{-n} S$, and the quotient field of R is the p-dimensional $Q(S)$ vector space

$$(\cup_n t^{-n} S) + y(\cup_n t^{-n} S) + \cdots + y^{p-1}(\cup_n t^{-n} S).$$

Every finitely generated R-submodule of it is also finitely generated as an S-module, and these are all contained in one of the modules

$$M_m := t^{-m}[S + yS + \cdots + y^{p-1}S].$$

On the other hand, for every n, we can write our equation as

$$\left(\frac{y - \sum_{i=1}^n x_i t^i}{t^{n+1}} \right)^p = \sum_{i \geq n+1} x_i^p t^{p(i-n-1)},$$

and this shows that

$$s_n := -\sum_{i=0}^{n} x_i t^{i-n-1} + t^{-n-1} y$$

is integral over R. Since $s_n \in M_m$ only if $n \leq m + 1$, the S-submodule generated by all the s_n cannot be finitely generated.   □

It is also instructive to see how the attempt to resolve this singularity using the blow-up process would work for R.

The maximal ideal of R is generated by (t, y), and we see that

$$R[\tfrac{y}{t}] = S[\tfrac{y}{t}]/((\tfrac{y}{t})^p - \sum_i x_i^p t^{pi-p}).$$

Here we should make the coordinate change $y_1 := \tfrac{y}{t} - x_1$ to get that

$$R[\tfrac{y}{t}] \cong S[y_1]/(y_1^p - \sum_{i\geq 1} x_{i+1}^p t^{pi}).$$

Aside from a reindexing of the variables, this is just like the original ring R, and from now on the process repeats forever.

EXAMPLE 1.105. (cf. [Nag62, App.Exmp.1]) Here we give an example of a Noetherian integral domain of dimension 1 with infinitely many singular points.

Let K be any field and $K[x_1, x_2, \ldots]$ the polynomial ring in infinitely many variables. Consider the subring $R := K[x_1^2, x_1^3, x_2^2, x_2^3, x_3^2, x_3^3, \ldots]$, where we have cusps along the hyperplanes $x_i = 0$. Set $P_i := (x_i^2, x_i^3)$ and $M := R \setminus (\cup_i P_i)$, and let $S := R_M$ be the localization of R at M.

The maximal ideals in S are $P_i S$, and the localizations are all isomorphic to an ordinary cusp $K'[y, z]/(y^2 - z^3)$, where K' is the rational function field of infinitely many variables over K.

Thus we see that every closed point is a singular point.

Each of these points can be resolved by a single blow-up, but we need infinitely many of them. The normalization of S is the localization of $K[x_1, x_2, \ldots]$ at the multiplicative set $K[x_1, x_2, \ldots] \setminus \cup_i (x_i)$. It is an easy exercise to check that S is Noetherian (cf. [AM69, Ch.11.Exrc.4]).

CHAPTER 2

Resolution for Surfaces

Surface singularities are much more complicated than curves, and resolving them is a rather subtle problem. It seems that surfaces are still
special enough to connect resolutions directly to various geometric properties but not general enough to predict which methods generalize to higher
dimensions.

One culprit is the existence of a minimal resolution, which is studied in
Section 2.2 after some introductory examples in Section 2.1. While minimal
resolutions provide a powerful tool for the study of surface singularities,
from the point of view of a general resolution method they have been a
distraction so far.

Three ways of resolving surface singularities are studied in this chapter.

Riemann's approach using topology (Section 1.2) reappears in Section
2.3 as Jung's method for surfaces. It shows that every surface is birational
to a surface with cyclic quotient singularities. These are classified and
resolved in Section 2.4.

The Albanese proof for curves (Section 1.3) is generalized to any variety
in Section 2.5. It shows that every n-dimensional variety is birational to a
variety whose singularities have multiplicity $\leq n!$. The relevant background
material on multiplicities is reviewed in Section 2.8. In the surface case we
have only double points left to deal with, and these are relatively easy to
resolve, at least when the characteristic is different from 2. This is done in
Section 2.6.

These two approaches should be viewed as but fragments of general
resolution methods that simplify the singularities but do not fully resolve
them. For surfaces the remaining problems can be settled, though we will
need a variety of techniques of algebraic geometry. Thus this chapter ends
up technically much more complicated than Chapter 1 or even Chapter 3.
It seems that, after all, the best resolution methods are the simplest ones.

The technical continuity between curves and the general results of
Chapter 3 is provided by the method of maximal contact and Weierstrass
equations. Here we use it, in Section 2.7, to prove embedded resolution
for surfaces in characteristic zero. The Weierstrass preparation theorem
allows one to choose local coordinate systems that are optimally adapted

to resolution. Such coordinate systems are needed only at the finitely many "worst" points, so there is no need to worry about their compatibility. In higher dimensions, compatibility becomes a serious problem, and one of the important steps in Chapter 3 is to eliminate the use of the preparation theorem.

2.1. Examples of resolutions

When resolving in practice, we frequently start with a surface sitting in a smooth variety and wish to understand how it transforms under a birational map of the ambient smooth variety.

DEFINITION 2.1. Let $g : X \dashrightarrow Y$ be a rational map that is defined on an open set $X^0 \subset X$, and let $Z \subset X$ be a subvariety such that $Z \cap X^0$ is dense in Z. Then the closure of $g(Z \cap X^0)$ is called the *birational transform* of Z and denoted by $g_*(Z)$. (In [Har77, Sec.II.7], this is called the *strict transform*.)

If $D = \sum m_i D_i$ is a divisor, its birational transform is $g_*(D) := \sum m_i g_*(D_i)$.

Frequently, g is the inverse of a birational map $f : Y \dashrightarrow X$, in which case the birational transform is denoted by $f_*^{-1}(Z) \subset Y$.

2.2 (Computing blow-ups). Let $S := (f(x, y, z) = 0) \subset \mathbb{A}^3$ be a surface singularity of multiplicity m, and let us consider what happens when we blow up the origin. The blow-up is covered by three affine charts, and we get the respective equations

$$
\begin{aligned}
x_1^{-m} f(x_1, y_1 x_1, z_1 x_1) = 0, &\quad \text{where } x = x_1, \ y = y_1 x_1, \ z = z_1 x_1, \\
y_2^{-m} f(x_2 y_2, y_2, z_2 y_2) = 0, &\quad \text{where } x = x_2 y_2, \ y = y_2, \ z = z_2 y_2, \\
z_3^{-m} f(x_3 z_3, y_3 z_3, z_3) = 0, &\quad \text{where } x = x_3 z_3, \ y = y_3 z_3, \ z = z_3.
\end{aligned}
$$

These equations define the *blow-up* or *birational transform* of S at the origin $0 \in \mathbb{A}^3$. We frequently write it as $B_0 S \subset B_0 \mathbb{A}^3$.

If we perform further blow-ups, we need double subscripts like x_{ij}, but the notation gets even worse if we have to blow up points that are not at the origin of any of these charts.

I find that these computations are perfectly manageable when done alone, where a certain sloppiness of notation is no problem, but they get hopelessly messy when I try to do them in public and to keep track of all the indices.

Instead of trying to name the new coordinates all the time, I sometimes write them down as rational functions in x, y, z. As long as we blow up only smooth subvarieties, all the local charts are surfaces in $\mathbb{A}^3$, so three coordinates and one equation are enough.

For example, the blow-up sequence for the curve singularity $x^3 = y^8$ can be represented as

$$x^3 = y^8 \mapsto \left(\tfrac{x}{y}\right)^3 = y^5 \mapsto \left(\tfrac{x}{y^2}\right)^3 = y^2 \mapsto \left(\tfrac{x}{y^2}\right) = \left(\tfrac{y^3}{x}\right)^2,$$

at which point we have the equation $x_3 = y_3^2$, where $x_3 = \tfrac{x}{y^2}$ and $y_3 = \tfrac{y^3}{x}$. Usually it is more useful to see the inverse of the coordinate change; here it is given by $x = x_3 y^2 = x_3^3 y_3^2$ and $y = y_3 x_3$.

It is best, however, to look at the examples in this section as tour guides only, not as complete proofs.

2.3 (Resolving $z^2 = f_m(x,y)$). Consider singularities of the form

$$S := \left(z^2 = f_m(x,y)\right) \subset \mathbb{A}_k^3,$$

where f_m is homogeneous of degree m without multiple roots.

We can view S as a double cover of the (x,y)-plane, branched along the curve $B := (f_m = 0)$. The branch curve B has an ordinary m-fold point at the origin, which can be resolved by a single blow-up. Let this be $\pi : B_0 \mathbb{A}^2 \to \mathbb{A}^2$ with exceptional curve $E \subset S$. Let $B' \subset B_0 \mathbb{A}^2$ denote the birational transform of B.

Correspondingly, let

$$S_1 \subset \mathbb{A}_z^1 \times B_0 \mathbb{A}^2$$

be the birational transform of S. Note that S_1 can be obtained from S by blowing up the ideal $(x,y) \subset \mathcal{O}_S$. (This is not the maximal ideal since it does not contain z.)

Over one of the standard local charts on $B_0 \mathbb{A}^2$, S_1 is given by an equation

$$z^2 = x_1^m f_m(1, y_1).$$

Thus we see that its normalization $\bar{S}_1$ is given by an equation

$$\begin{aligned} \bar{S}_1 &:= \left(z_2^2 = f_m(1, y_1)\right) && \text{if } m \text{ is even, and} \\ \bar{S}_1 &:= \left(z_2^2 = x_1 f_m(1, y_1)\right) && \text{if } m \text{ is odd,} \end{aligned}$$

where $z_2 = z x_1^{-\lfloor m/2 \rfloor}$. The blow-ups can be summarized as

$$z^2 = f_m(x,y) \mapsto z^2 = x^m f_m\left(1, \tfrac{y}{x}\right) \mapsto \left(\tfrac{z}{x^{\lfloor m/2 \rfloor}}\right)^2 = x^\epsilon f_m\left(1, \tfrac{y}{x}\right),$$

where $\epsilon = 0$ if m is even and $\epsilon = 1$ if m is odd.

We can also obtain these by first blowing up the origin in $\mathbb{A}^3$ to get (in an affine chart)

$$\left(\tfrac{z}{x}\right)^2 = x^{m-2} f_m\left(1, \tfrac{y}{x}\right)$$

and then normalizing.

From now on we have to treat two cases separately, depending on the parity of m.

Even case. If m is even, then $\bar{S}_1$ is smooth and the composite $f : \bar{S}_1 \to S$ is a resolution.

The exceptional curve E_S of $\bar{S}_1 \to S$ is a double cover of the exceptional curve $E \subset B_0 \mathbb{A}^2$, ramified at the intersection points $E \cap B'$. Thus E_S is hyperelliptic of genus $\frac{1}{2}m - 1$. Since E_S is the pullback of $E \subset B_0 \mathbb{A}^2$ by a double cover, we conclude that $E_S^2 = -2$.

The canonical class $K_{\bar{S}_1}$ can be computed two different ways.

We know that $K_{\bar{S}_1} = f^* K_S + (\text{exceptional curves})$. Here there is only one exceptional curve E_S, and so $K_{\bar{S}_1} = f^* K_S - a E_S$ and the coefficient of E_S can be computed using the adjunction formula

$$2 p_a(E_S) - 2 = \big(E_S \cdot (E_S + K_{\bar{S}_1}) \big).$$

We get that $2\big(\frac{1}{2}m - 1\big) - 2 = -2(1 - a)$, which gives $a = \frac{1}{2}m - 1$.

I usually find it easier to work out these coefficients from the Jacobian of f. Assume for simplicity that char $k \neq 2$.

A local generator of K_S is given by

$$\frac{dx \wedge dy}{\partial(z^2 - f_m)/\partial z} = \frac{dx \wedge dy}{2z}.$$

Similarly, a local generator of $K_{\bar{S}_1}$ is given by

$$\frac{dx_1 \wedge dy_1}{2 z_2}.$$

Since $x = x_1, y = y_1 x_1, z = z_2 x_1^{m/2}$, we conclude that

$$f^* \frac{dx \wedge dy}{2z} = x_1^{1 - \frac{m}{2}} \cdot \frac{dx_1 \wedge dy_1}{2 z_2}.$$

Thus we get a pole of multiplicity $\frac{m}{2} - 1$ along E_S, and so

$$K_{\bar{S}_1} = f^* K_S - \big(\tfrac{m}{2} - 1 \big) E_S.$$

Odd case. If m is odd, then $\bar{S}_1$ has m ordinary double points at the intersection points of E and B'. The exceptional curve E_S of $\bar{S}_1 \to S$ is isomorphic to $E \cong \mathbb{P}^1$. Since $2 E_S$ is the pullback of $E \subset B_0 \mathbb{A}^2$ by a double cover, we conclude that $E_S^2 = -\frac{1}{2}$.

We can resolve each of the double points by a single blow-up to obtain the smooth surface $g : S_2 \to \bar{S}_1 \to S$. The exceptional curves are the birational transform of E_S (again denoted by E_S) and m new rational curves $C_1, \ldots, C_m$ with self-intersection -2. Thus we see that

$$K_{S_3} = f^* K_S - a E_S - b \sum C_i,$$

and a computation as in the first case gives that $a = \frac{m-3}{2}$.

In order to compute b, after a coordinate change, we can assume for simplicity that $f_m(x, y) = y g(x, y)$. Then the relevant sequence of blow-ups is

$$z^2 = yg(x,y) \;\mapsto\; z^2 = x^m \tfrac{y}{x} g\left(1, \tfrac{y}{x}\right) \;\mapsto\; \left(\tfrac{z}{x^{(m-1)/2}}\right)^2 = x \tfrac{y}{x} g\left(1, \tfrac{y}{x}\right)$$

$$\mapsto \left(\tfrac{z}{x^{(m+1)/2}}\right)^2 = \tfrac{y}{x^2} g\left(1, x\tfrac{y}{x^2}\right).$$

Thus at the end we have

$$z_3^2 = y_3 g(1, x_3 y_3) \quad \text{and} \quad x = x_3,\; y = y_3 x_3^2,\; z = z_3 x_3^{(m+1)/2}.$$

Therefore,

$$f^* \frac{dx \wedge dy}{2z} = x_3^{-(m-3)/2} \cdot \frac{dx_3 \wedge dy_3}{2z_3}.$$

Thus we see that

$$K_{S_3} = f^* K_S - \left(\tfrac{m-3}{2}\right) E_S - \left(\tfrac{m-3}{2}\right) \sum C_i.$$

Note. Instead of blowing up each singular point of $\bar{S}_1$ separately, it may be easier to resolve $\bar{S}_1$ by blowing up E_S. (On a smooth surface we do not change anything by blowing up a curve. More generally, nothing changes if we blow up Cartier divisors. In our case, however, $\bar{S}_1$ is singular, and E_S is not Cartier at the singular points.) We get two charts; the more interesting one is

$$z^2 = f(x,y) \;\mapsto\; z^2 = x^m f\left(1, \tfrac{y}{x}\right) \;\mapsto\; \left(\tfrac{z}{x^{(m-1)/2}}\right)^2 = x f\left(1, \tfrac{y}{x}\right)$$

$$\mapsto \left(\tfrac{z}{x^{(m+1)/2}}\right)^2 x = f\left(1, x\tfrac{y}{x}\right).$$

At the end we have

$$z_3^2 x_3 = f_m(1, y_3), \quad \text{where } x = x_3,\; y = x_3 y_3,\; z = z_3 x_3^{(m+1)/2}.$$

2.4 (Resolving $x^2 + y^3 + z^6 = 0$). Assume that char $k \neq 2, 3$. After one blow-up, we get one singular point with equation

$$S_1 := (x_1^2 + y_1^3 z_1 + z_1^4 = 0), \quad \text{where } x = x_1 z_1,\; y = y_1 z_1,\; z = z_1.$$

The exceptional curve is $C = (z_1 = x_1 = 0)$, and hence it is a smooth rational curve. Furthermore,

$$f_1^* \frac{dy \wedge dz}{2x} = \frac{dy_1 \wedge dz_1}{2x_1}$$

shows that $K_{S_1} = f_1^* K_S$. The singularity of S_1 is of the type studied in (2.3) for $m = 4$. Thus we know how to construct a resolution $f_2 : S_2 \to S_1$.

Putting the information together, we obtain a resolution $f : S_2 \to S$ with two exceptional curves $C, E \subset S_2$ where C is smooth and rational, E is smooth and elliptic with $(E^2) = -2$ and the two curves intersect transversally at a single point. Moreover,

$$K_{S_2} = f_2^* K_{S_1} - E = f^* K_S - E.$$

This implies that $(C \cdot K_{S_2}) = -1$, and thus $C \subset S_2$ is a -1-curve. In order to get the minimal resolution (2.16), we need to contract C as in (2.14).

Hence the minimal resolution is $f_{\min} : S_{\min} \to S$ with a single exceptional curve $E \subset S_{\min}$, which is smooth and elliptic with $(E^2) = -1$.

2.5 (Resolving $x^2 + y^3 + z^7 = 0$). We leave it to the reader to check that this singularity can be resolved the same way as $x^2 + y^3 + z^6 = 0$. We need two blow-ups and a normalization. Again the resolution is nonminimal. The minimal resolution is $f_{\min} : S_{\min} \to S$ with a single exceptional curve $E \subset S_{\min}$, which is a rational curve with a single cusp and $(E^2) = -1$.

2.6 (Dual graphs). It is frequently very convenient to represent a reducible curve $C = \cup C_i$ on a surface by a graph whose vertices are the irreducible components of C and two vertices are connected by an edge iff the corresponding curves intersect. Depending on what one has in mind, extra information is frequently added to the graph. This may include the self-intersection number or the genus of the curve C_i for each vertex and the intersection number $(C_i \cdot C_j)$ for each edge. This graph, with various extra information added, is called the *dual graph* of the reducible curve $C = \cup C_i$.

This description has been especially useful when $0 \in S$ is a normal surface singularity with resolution $f : S' \to S$ and $\cup C_i = f^{-1}(0) \subset S'$ is the exceptional curve. In many examples we get dual graphs with few edges, so the picture is rather transparent. (On the other hand, for $S :=$ $(x^m + y^m + z^{m+1} = 0)$, the dual graph of the minimal resolution—obtained by blowing up the origin—is the complete graph on m vertices, so the dual graph is not always the best way to describe a resolution.)

We see in (2.12) that the self-intersection numbers $(C_i \cdot C_i)$ are always negative, and usually one uses the number $-(C_i \cdot C_i)$ to represent a vertex. In our examples most of the exceptional curves C_i are smooth and rational. I put a box around the number if the curve is not smooth or not rational.

For example, the resolutions constructed in (2.3) are represented by the graph $\boxed{2}$ if $m \geq 4$ is even (here we would clearly benefit from adding the genus to the notation), and for $m = 3$ and $m = 5$ we get

$$
\begin{array}{ccc}
2 & & 2 \quad 2 \quad 2 \\
| & \text{and} & \diagdown \ | \ \diagup \\
2 \ - \ 2 \ - \ 2 & & 2 \ - \ 3 \ - \ 2
\end{array}
$$

EXAMPLE 2.7 (Du Val singularities). The reader may enjoy working out the resolutions of the *Du Val* singularities. They are also called *rational double points* of surfaces or *simple* surface singularities. These are the only normal surface singularities $0 \in S$ with a resolution $f : S' \to S$ such that $K_{S'} = f^* K_S$. See [KM98, Sec.4.2] or [Dur79] for more information. (The equations below are correct in characteristic zero. The dual graphs are correct in every characteristic.)

A_n: $x^2 + y^2 + z^{n+1} = 0$, with $n \geq 1$ curves in the dual graph:

$$2 \ - \ 2 \ - \ \cdots \ - \ 2 \ - \ 2$$

D_n: $x^2 + y^2 z + z^{n-1} = 0$, with $n \geq 4$ curves in the dual graph:

$$
\begin{array}{ccccccc}
 & & 2 & & & & \\
 & & | & & & & \\
2 & - & 2 & - & \cdots & - & 2 & - & 2
\end{array}
$$

E_6: $x^2 + y^3 + z^4 = 0$, with six curves in the dual graph:

$$
\begin{array}{ccccccccc}
 & & & & 2 & & & & \\
 & & & & | & & & & \\
2 & - & 2 & - & 2 & - & 2 & - & 2
\end{array}
$$

E_7: $x^2 + y^3 + yz^3 = 0$, with seven curves in the dual graph:

$$
\begin{array}{ccccccccccc}
 & & & & 2 & & & & & & \\
 & & & & | & & & & & & \\
2 & - & 2 & - & 2 & - & 2 & - & 2 & - & 2
\end{array}
$$

E_8: $x^2 + y^3 + z^5 = 0$, with eight curves in the dual graph:

$$
\begin{array}{ccccccccccccc}
 & & & & 2 & & & & & & & & \\
 & & & & | & & & & & & & & \\
2 & - & 2 & - & 2 & - & 2 & - & 2 & - & 2 & - & 2
\end{array}
$$

2.8 (Resolving in practice). There seems to be a very wide gulf between our theoretical understanding of resolutions and practical computations. As (2.4) shows, for a surface singularity, various resolution algorithms may give a nonminimal resolution. In fact, this happens in many cases.

For concrete surfaces, it is frequently easier to use some special features to get a quick and efficient way of resolving them.

For instance, one notices right away that the equation $x^2 + y^3 + z^6 = 0$ becomes homogeneous if we declare that $\deg x = 3$, $\deg y = 2$ and $\deg z = 1$. Such equations are called *weighted homogeneous*.

In order to resolve weighted homogeneous varieties, it is almost always better to take the weights into account. This approach is best codified using *toric geometry* and *toric blow-ups* [Oda88, Ful93].

2.2. The minimal resolution

Before we start resolving in earnest, we assume that resolutions exist, and we study the relationship between the various resolutions of a surface. We investigate birational morphisms between surfaces and show that there is a minimal resolution.

We start with some auxiliary results on birational maps in general.

LEMMA 2.9. *Let $f : X \to Y$ be a projective and generically finite morphism, X, Y normal and quasi-projective. Then there is an effective Cartier divisor W on X such that $-W$ is f-ample.*

Proof. Let H be an effective, ample divisor on X. Its push-forward $f(H) \subset Y$ is a Weil divisor on Y; it is thus contained in an effective Cartier divisor D since Y is quasi-projective. Thus $W := f^*(D) - H$ is effective, and if $C \subset X$ is any curve contracted by f, then $(W \cdot C) = (f^*D \cdot C) - (H \cdot C) = -(H \cdot C)$, so $-W$ is f-ample.   □

COMPLEMENT 2.10. *One can choose W to be f-exceptional in the following cases:*

(1) *f is birational and Y is smooth (or $\mathbb{Q}$-factorial),*
(2) *f is birational, X is smooth and $\dim X = 2$.*

Proof. If f is birational and Y is smooth, then we can take $D = f(H)$ and so $W = H - f^*(f(D))$ is f-exceptional. Note that this argument only needs that $f^*(f(D))$ makes sense as a cycle with rational coefficients, which is the case if $mf(H)$ is Cartier for some $m > 0$. This holds if Y is $\mathbb{Q}$-factorial.

If $\dim X = 2$ and X is smooth, write $W = W' + F$, where W' consists of all f-exceptional components and F the rest. F is a Cartier divisor if X is smooth. If $\dim X = 2$, then f does not contain any exceptional curve $C \subset X$, and thus $(F \cdot C)$ is nonnegative. Hence, W' is also f-ample and consists of exceptional divisors only.

See (2.19) for comments on the existence of such W' in general.   □

COROLLARY 2.11. *Let $f : X \to Y$ be a projective and generically finite morphism, X, Y normal and quasi-projective. Let Z be an exceptional Cartier divisor such that $-Z$ is not effective. Then there is an exceptional curve C such that $(C \cdot Z) < 0$.*

Proof. Choose the maximal $a \in \mathbb{Q}$ such that $aZ \leq W$, where W is as in (2.9). Then $W - aZ$ is effective, and there is an exceptional curve C that is not contained in $\mathrm{Supp}(W - aZ)$. Then $a(C \cdot Z) = (C \cdot W) - (C \cdot (W - aZ)) < 0$.   □

THEOREM 2.12 (Hodge index theorem). *Let $f : X \to Y$ be a projective and generically finite morphism, Y quasi-projective and X a smooth surface. Then the intersection form $(C_i \cdot C_j)$ is negative-definite on exceptional curves.*

Proof. Consider the function $f(x_1, \ldots, x_m) = (\sum x_i C_i \cdot \sum x_i C_i)$. It is enough to prove that, in any cube $0 \leq x_i \leq N$, f has a strict maximum at the origin. Since

$$\tfrac{\partial f}{\partial x_j}(Z) = 2(C_i \cdot Z),$$

we see from (2.11) that f increases as we move toward the origin parallel to one of the coordinate axes.   □

THEOREM 2.13 (Factorization of birational morphisms). *Let $f : X \to Y$ be a projective, birational morphism between smooth surfaces. Then f is the composite of blowing up smooth points.*

Proof. The traditional proof (cf. [Sha94, Sec.IV.3]) starts with a point $y \in Y$, where f^{-1} is not defined and proves that f can be factored through its blow-up $X \to B_p Y \to Y$. Thus we get the factorization by starting at the bottom.

Here is another proof, in the spirit of Mori's program, which starts at the top. (See, for instance, [KM98] for an introduction to Mori's program, also called the minimal model program.)

Differential forms can be pulled back, and thus there is map $f^* \omega_Y \to \omega_X$. Written in terms of divisors, we get that $K_X \sim f^* K_Y + E$, where E is effective and f-exceptional.

By (2.11) there is an exceptional curve E_1 such that $(E_1 \cdot E) < 0$. From (2.12) we conclude that $(E_1 \cdot E_1) < 0$. The adjunction formula now gives

$$2p_a(E_1) - 2 = (E_1 \cdot K_X + E_1) = (E_1 \cdot E) + (E_1 \cdot E_1) < 0.$$

Thus $p_a(E_1) = 0$, $E_1 \cong \mathbb{P}^1$ and $(E_1 \cdot E) = (E_1 \cdot E_1) = -1$. Thus E_1 is contractible to a smooth point by (2.14). □

THEOREM 2.14 (Castelnuovo's contractibility criterion). *Let X be a smooth projective surface and $\mathbb{P}^1 \cong E \subset X$ a curve such that $(E \cdot E) = -1$.*

Then there is a smooth projective pointed surface $y \in Y$ and a birational morphism $f : X \to Y$ such that $f(E) = y$ and $f : X \setminus E \to Y \setminus \{y\}$ is an isomorphism.

Proof. Pick a very ample divisor H on X such that $H^1(X, \mathcal{O}_X(nH)) = 0$ for $n \geq 1$. Using the sequences

$$0 \to \mathcal{O}_X(nH + iE) \to \mathcal{O}_X(nH + (i+1)E) \to \mathcal{O}_X(nH + (i+1)E)|_E \to 0$$

we see that

(1) $|nH + iE|$ is very ample on $X \setminus E$ for $n, i \geq 0$,
(2) $H^1(X, \mathcal{O}_X(nH + iE)) = 0$ if $((nH + iE) \cdot E) \geq -1$,
(3) $H^0(X, \mathcal{O}_X(nH + iE)) \twoheadrightarrow H^0(E, \mathcal{O}_E((nH + iE)|_E))$ is onto if $((nH + iE) \cdot E) \geq 0$, and
(4) $|nH + iE|$ is base point free if $((nH + iE) \cdot E) \geq 0$.

Set $m = (H \cdot E)$. Then $(H + mE \cdot E) = 0$ and $(H + (m-1)E \cdot E) = 1$, and hence by (3) there is a curve $D \in |H + (m-1)E|$ that intersects E transversally in a single point. By (4) we can choose $D' \in |H + mE|$ disjoint from E. From the sequence

$$\mathcal{O}_X((n-1)(H + mE) + E) \hookrightarrow \mathcal{O}_X(n(H + mE)) \twoheadrightarrow \mathcal{O}_D(n(H + mE)|_D)$$

we conclude that $H^0(X, \mathcal{O}_X(n(H + mE))) \twoheadrightarrow H^0(D, \mathcal{O}_D(n(H + mE)|_D))$ is onto. Moreover, $n(H + mE)|_D$ is very ample for $n \gg 1$.

Let $f : X \to Y$ be the morphism given by $|n(H+mE)|$ for some $n \gg 1$. As we saw, f contracts E to a point y, $|n(H+mE)|$ is very ample on $X \setminus E$, and $|n(H + mE)|$ is also very ample on D.

Thus $f(D + (n-1)D') = f(C + D + (n-1)D') \subset Y$ is a curve, which is smooth at y. It is also a hyperplane section of Y since $C + D + (n-1)D' \in |n(H + mE)|$, and thus Y is smooth at y. $\square$

REMARK 2.15. It is easy to modify the above proof to show that any curve $\mathbb{P}^1 \cong E \subset X$ such that $(E \cdot E) < 0$ can be contracted to a point, though for $(E \cdot E) \leq -2$ the resulting point is singular.

THEOREM 2.16 (Minimal resolutions). *Let Y be a surface. Then there is a unique birational morphism $f : X \to Y$ from a smooth surface X to Y, which is characterized by either of the following properties.*

(1) *Every resolution $g : Y' \to Y$ can be factored as $g : Y' \xrightarrow{\sigma} X \xrightarrow{f} Y$.*
(2) *K_X is f-nef, that is, $(K_X \cdot E) \geq 0$ for every f-exceptional curve $E \subset X$.*

Proof. For now we prove this under the additional assumption that there is at least one resolution $g : S \to Y$. Once we prove that resolutions exist, we will have a complete proof of (2.16).

Let $g : S \to Y$ be a resolution. If K_S is not f-nef, there is an exceptional curve $E \subset X$ such that $(E \cdot K_X) < 0$. $(E \cdot E) < 0$ by (2.12); hence, using the adjunction formula as in the proof of (2.13) we conclude that E is a -1-curve and it can be contracted by (2.14). As the number of exceptional curves is finite, we eventually get a resolution $f : X \to Y$ such that K_X is f-nef.

In order to prove uniqueness, let $f_i : X_i \to Y$ be two resolutions such that K_{X_i} is f_i-nef. Let S be the normalization of the closure of the graph of the birational map between X_1 and X_2 with birational projections $g_i : S \to X_i$.

(One could take instead a resolution of S and use the factorization theorem (2.13) to get our result. Our aim here is to give a proof that does not use resolution and generalizes well to higher dimensions.)

Let $s \in S$ be a smooth point with local coordinates z_1, z_2 and x_i, y_i local coordinates at $g_i(s) \in X_i$. Then

$$g_i^*(dx_i \wedge dy_i) = \mathrm{Jac}\big(\tfrac{x_i, y_i}{z_1, z_2}\big) \cdot dz_1 \wedge dz_2.$$

Thus we can write $K_S = g_i^* K_{X_i} + E_i$, where E_i is effective and its support is the whole g_i-exceptional locus. Write $E_1 - E_2 = F_1 - F_2$, where F_1 and F_2 have no components in common. Here K_S need not be Cartier, but we get an equality of Cartier divisors

$$E_1 - E_2 = F_1 - F_2 = g_2^* K_{X_2} - g_1^* K_{X_1}.$$

If $F_1 \neq 0$, then by (2.11) there is a g_1-exceptional curve C such that $C \cdot (F_1 - F_2) < 0$.

On the other hand, since $g_{1*}C = 0$,

$$\big(C \cdot (g_2^* K_{X_2} - g_1^* K_{X_1})\big) = (g_{2*}C \cdot K_{X_2}) - (g_{1*}C \cdot K_{X_1}) = (g_{2*}C \cdot K_{X_2}) \geq 0.$$

This is a contradiction. Thus $F_1 = 0$ and similarly $F_2 = 0$; hence $E_1 = E_2$.

Since S is the normalization of the closure of the graph of the birational map between the X_i, no curve in S can be exceptional for both g_1 and g_2. Thus the g_i are both isomorphisms. $\qquad\square$

ASIDE 2.17. In higher dimensions, the proof of (2.16) gives the following. (See [KM98, Sec.2.3 and Sec.3.8] for the definition of canonical, terminal and related results.)

THEOREM. *Let $f_i : X_i \to Y$ be two proper morphisms such that K_{X_i} is f_i-nef and the X_i have canonical singularities. Let $g_i : X \to X_i$ be proper birational morphisms. Then $g_1^* K_{X_1} = g_2^* K_{X_2}$.*

If, moreover, X_1 has terminal singularities, then every g_1-exceptional divisor is also g_2-exceptional.

For any normal surface, the construction of a resolution is local around the singular points in the strongest possible sense.

PROPOSITION 2.18. *Let S be a normal surface with singular points $p_1, \ldots, p_n \in S$. There is a one-to-one correspondence between*

(1) *resolutions $f : X \to S$ and*
(2) *resolutions $f_i : \hat{X}_i \to \hat{S}_{p_i}$ for every i, where $\hat{S}_{p_i} := \operatorname{Spec} \widehat{\mathcal{O}}_{p_i, S}$ is the completion of S at p_i.*

Proof. If $f : X \to S$ is a resolution, then by base change we get resolutions of the completions $f_i : \hat{X}_i \to \hat{S}_{p_i}$.

Conversely, assume that we have resolutions $f_i : \hat{X}_i \to \hat{S}_{p_i}$ for every i. By (2.10.2) and (2.19.1) we can get $\hat{X}_i$ by blowing up an ideal $\hat{J}_i \subset \hat{\mathcal{O}}_{S_{p_i}}$ such that $\hat{\mathcal{O}}_{S_{p_i}} / \hat{J}_i$ has finite length. All such ideals come from an ideal $J_i' \subset \mathcal{O}_{S_{p_i}}$, and these in turn come from ideal sheaves $J_i \subset \mathcal{O}_S$ whose cosupport is at p_i. Take now $J := \cap J_i$. Then $f : B_J S \to S$ is a projective, birational morphism, which is an isomorphism above $S \setminus \{p_1, \ldots, p_n\}$, and over $\hat{S}_{p_i}$ it gives the original resolutions $f_i : \hat{X}_i \to \hat{S}_{p_i}$. Thus X is nonsingular, and so $f : X := B_J S \to S$ is the required resolution. $\qquad\square$

2.19 (Antiample exceptional divisors). In (2.9), if f is birational and Y is smooth, then we can take $D = f(H)$ to obtain an f-exceptional divisor W such that $-W$ is f-ample. In general, however, no f-exceptional divisor is relatively antiample. This fails even for surfaces. Let $Y = (x^3 + y^3 + z^3 = 0) \subset \mathbb{A}^3$ be an elliptic cone. Blow up the origin to get a resolution

$f_1 : Y_1 := B_0 Y \to Y$ whose exceptional divisor is the elliptic curve $E = (x^3 + y^3 + z^3 = 0) \subset \mathbb{P}^2$. Pick a point $p \in E \subset B_0 Y$ such that $\mathcal{O}_{\mathbb{P}^2}(1)|_E(-3p)$ is not torsion in the Picard group of E. Blow up p to get $Y_2 := B_p Y_1$ with new exceptional curve C. Then contract the birational transform of E to get $h : Y_2 \to X$ and $g : X \to Y$. g has one exceptional divisor, which is a smooth rational curve C_X, and no multiple of C_X is Cartier. Indeed, if $m C_X$ is Cartier, then $h^*(m C_X) \sim mC + nE$ can be moved away from E, and thus $\mathcal{O}_{Y_2}(mC + nE)|_E \cong \mathcal{O}_E$. Thus $\mathcal{O}_E(-mp)$ is isomorphic to $\mathcal{O}_{Y_2}(nE)|_E \cong \mathcal{O}_{\mathbb{P}^2}(-n)|_E$. By our assumption this implies $m = n = 0$, and so there are no f-exceptional Cartier divisors.

In general, we have the following characterization.

Claim 2.19.1. Let $f : X \to Y$ be a projective, birational morphism. The following are equivalent.

   (i) There is an exceptional divisor W on X such that $-W$ is f-ample.
   (ii) X can be obtained from Y by blowing up a subscheme $Z \subset Y$ of codimension at least 2.
   (iii) There is a sequence of blow-ups $X = Y_n \to \cdots \to Y_0 = Y$ where each $f_{i+1} : Y_{i+1} \to Y_i$ is obtained by blowing up a subscheme $Z_i \subset Y_i$ of codimension at least 2.

Proof. If $X = B_Z Y$, then $f^{-1}(I_Z)$ gives an f-ample locally free ideal sheaf on X. Thus $f^{-1}(I_Z) = \mathcal{O}_X(-W)$ for some Cartier divisor W. Since $f(W) \subset \operatorname{Supp} Z$, W is f-exceptional.

Conversely, f is the blow-up of $f_* \mathcal{O}_X(-eW)$ for $e \gg 1$ (cf. [Har77, II.7.17.Step 2]), and so take $\mathcal{O}_Z = \mathcal{O}_Y / f_* \mathcal{O}_X(-eW)$.

Finally, assume that we have $f_i : Y_i \to Y_{i-1}$ and effective f_i-exceptional divisors $W_i \subset Y_i$ such that $-W_i$ is f_i-ample. Then $W_2 + c_1(f_2^* W_1)$ is $f_1 \circ f_2$-exceptional and $-(W_2 + c_1(f_2^* W_1))$ is $f_1 \circ f_2$-ample for $c_1 \gg 1$. Repeating this we eventually get a Cartier divisor $W \subset X$, which is exceptional over Y and such that $-W$ is relatively ample. □

ASIDE 2.20 (Geometric genus of a surface singularity). As an illustration on how to use resolutions in the study of singularities, we define the geometric genus of surface singularities and derive the basic duality that is central to its usefulness.

This is the only part of these notes where we seriously use the cohomology theory of varieties and some of its harder results presented in [Har77, Chap.III].

Foundational comments. Although [Har77, Sec.III.7] defines ω_X only for projective schemes, by restriction it can be defined on any quasi-projective scheme. There is a possible problem of independence of the compactification; this can be settled using (3.39). We also need that ω_X is reflexive, which is easy to derive from the basic definition. For X projective, let

$\omega_X \hookrightarrow \bar{\omega}_X$ be the reflexive hull. Then $H^n(X, \bar{\omega}_X) = H^n(X, \omega_X) = k$ since $\bar{\omega}_X / \omega_X$ is supported in dimension $< n$. This gives an inverse $\bar{\omega}_X \to \omega_X$ of the injection, and thus $\bar{\omega}_X = \omega_X$.

We start with the 2-dimensional Grauert-Riemenschneider vanishing theorem, which is the only case that holds in positive characteristic. (See [KM98, Sec.2.5] for generalizations to higher dimensions.)

Theorem 2.20.1. Let $f : X \to Y$ be a projective and generically finite morphism, Y normal and quasi-projective and X a smooth surface. Then $R^1 f_* \omega_X = 0$.

Proof. Pick any 1-cycle Z whose support is the whole exceptional set. By the theorem on formal functions (cf. [Har77, Sec.III.11])

$$R^1 f_* \omega_X = \varprojlim H^1(mZ, \omega_X|_{mZ}).$$

Thus it is enough to prove that $H^1(mZ, \omega_X|_{mZ}) = 0$ for every $m \geq 1$. Rewrite the restriction as

$$\omega_X|_{mZ} = \omega_X(mZ)(-mZ)|_{mZ} = \omega_{mZ} \otimes \mathcal{O}_X(-mZ).$$

Note also that $\mathcal{O}_X(mZ)|_{mZ} = \mathcal{O}_X(mZ)/\mathcal{O}_X$. By Serre duality on mZ (cf. [Har77, Sec.III.7]), we thus need to prove that

$$H^0(mZ, \mathcal{O}_X(mZ)|_{mZ}) = H^0(mZ, \mathcal{O}_X(mZ)/\mathcal{O}_X) = H^0(X, \mathcal{O}_X(mZ)/\mathcal{O}_X)$$

vanishes. The latter is (2.20.2). $\qquad\qquad\square$

Lemma 2.20.2. With the above notation, let Z be an effective exceptional cycle. Then $H^0(X, \mathcal{O}_X(mZ)/\mathcal{O}_X) = 0$.

Proof. By (2.11) there is an exceptional curve C such that $(C \cdot Z) < 0$. Then $C \subset \text{Supp } Z$, and we have an exact sequence

$$0 \to \mathcal{O}_X(Z - C)/\mathcal{O}_X \to \mathcal{O}_X(Z)/\mathcal{O}_X \to \mathcal{O}_X(Z)/\mathcal{O}_X(Z - C) \to 0.$$

Since $\mathcal{O}_X(Z)/\mathcal{O}_X(Z - C) \cong \mathcal{O}_X(Z)|_C$ is a line bundle of negative degree on a reduced curve C, $H^0(X, \mathcal{O}_X(Z)/\mathcal{O}_X(Z - C)) = 0$ and we are done by induction. $\qquad\qquad\square$

Corollary 2.20.3. Let $f : X \to Y$ be a projective and birational morphism, and let X, Y be quasi-projective and normal surfaces. Assume that $R^1 f_* \omega_X = 0$. Then $\omega_Y / f_* \omega_X$ is dual to $R^1 f_* \mathcal{O}_X$.

Proof. Note that ω_Y is reflexive, and so $f_* \omega_X$ is naturally a subsheaf of ω_Y. The two agree except possibly at the image of the exceptional set of f. In particular, $f_* \omega_X(Z) = \omega_Y$ whenever Z is a sufficiently large f-exceptional cycle. For any such Z, the exact sequence

$$0 \to \omega_X \to \omega_X(Z) \to \omega_Z \to 0$$

gives

$$0 \to f_* \omega_X \to \omega_Y \to H^0(Z, \omega_Z) \to R^1 f_* \omega_X = 0.$$

By duality on Z, $H^0(Z, \omega_Z)$ is dual to $H^1(Z, \mathcal{O}_Z)$ and $R^1 f_* \mathcal{O}_X = H^1(Z, \mathcal{O}_Z)$ for Z sufficiently large, again by the theorem on formal functions. Thus

$$0 \to f_* \omega_X \to \omega_Y \to (R^1 f_* \mathcal{O}_X)^* \to 0$$

is exact, where $(\ )^*$ denotes the vector space dual. $\square$

Definition 2.20.4. Let $y \in Y$ be a normal surface singularity and $f : X \to Y$ a resolution of singularities. Then

$$\text{length}(\omega_Y / f_* \omega_X) = \text{length } R^1 f_* \mathcal{O}_X$$

is independent of X and is called the *geometric genus* of $y \in Y$.

In order to see independence, let $f' : X' \to Y$ be another resolution of singularities. As in [Har77, II.8.19] every global section of $\omega_{X'}$ pulls back to a global section of ω_X (even though $f'^{-1} \circ f$ need not be a morphism). Thus $f_* \omega_X = f'_* \omega_{X'}$.

2.3. The Jungian method

Jung's approach to resolution, introduced in [Jun08] and fully developed in [Wal35, Hir53] shows that strong embedded resolution for curves implies resolution for surfaces.

It can be generalized to show that strong embedded resolution for $(n-1)$-dimensional varieties implies that every n-dimensional variety X is birational to a variety X' with abelian quotient singularities (2.22).

In its original form, there were two problems.

First, in dimension n we get resolution, but we need *embedded* resolution to continue to higher dimensions. This obstacle was overcome by [BP96, Par99].

Second, one still has to deal with abelian quotient singularities. Resolution of abelian quotient singularities is understood locally using toric methods (see [Oda88, Ful93]), and toroidal methods allow one to patch the local resolutions together, see [Par99, KKMSD73].

Both of these would lead us too far, so we focus on the surface case. We need strong embedded resolution for curves; this was proved in (1.47). We classify and resolve abelian quotient singularities in Section 2.4 and as shown in (2.18), there is no patching problem in dimension 2.

2.21 (Jung's method). Let k be a field of characteristic zero and $Z \subset \mathbb{P}^N_k$ a projective surface. We intend to resolve its singularities in four steps.

(1) By repeatedly projecting from points outside Z, we get a finite and dominant morphism $\pi : Z \to \mathbb{P}^2$. Let $B \subset \mathbb{P}^2$ be the branch locus of π.

(2) Apply strong embedded resolution (which we already know by (1.47)) to the pair $B \subset \mathbb{P}^2$. Thus we obtain a smooth surface Y and a birational morphism $h : Y \to \mathbb{P}^2$ such that $h^{-1}(B)$ is a

simple normal crossing divisor. Let X be the normalization of Y in the rational function field $k(Z)$ and $g : X \to Y$ the corresponding morphism.

(3) We prove that X has very special, namely, *abelian quotient* singularities (2.22).

(4) Finally, we need to resolve abelian quotient singularities of surfaces. In (2.25.2) we show that this is equivalent to resolving singularities of the form $\mathbb{A}^2/G$, where G acts linearly. In the abelian case, explicit resolutions are constructed in (2.32).

DEFINITION 2.22. Let X be a normal k-variety. We say that X is the *quotient* of the normal variety Z by the finite group G if there is a Galois extension $K/k(X)$ with Galois group G such that the normalization of X in K is Z. We write this as $X = Z/G$. G also acts on the ring of regular functions $k[Z]$. Let $k[Z]^G$ denote the *ring of invariants*.

Claim 2.22.1. Let the notation be as above. Z is affine iff $X = Z/G$ is, and then $k[X] = k[Z]^G$.

Proof. Since $Z \to X$ is finite and surjective, Z is affine iff X is (cf. [Har77, III.Exrc.4.2]). Set $|G| = m$ and let $k[Z] = k[x_1, \dots, x_n]$. For any j, let $\sigma_i(x_j), i = 1, \dots, m$, be the elementary symmetric polynomials of the $\{g(x_j) : g \in G\}$. Then

$$k[\sigma_i(x_j)] \subset k[x_1, \dots, x_n]^G \subset k[x_1, \dots, x_n]$$

is finitely generated. Moreover, each x_j is integral over $k[\sigma_i(x_j)]$ as it satisfies the equation

$$x_j^m + \sum_i (-1)^i \sigma_i(x_j) x_j^{m-i} = 0.$$

Thus $k[x_1, \dots, x_n]$ is a finite $k[\sigma_i(x_j)]$-module; hence, $k[x_1, \dots, x_n]^G$ is also a finite $k[\sigma_i(x_j)]$-module. Therefore, it is also finitely generated as a ring. $\square$

We say that X has (algebraic) *quotient singularities* if there is an open affine cover $X = \cup X_i$ such that each X_i is the quotient of a smooth affine variety Z_i by a finite group G_i. If all the G_i are abelian, we say that X has *abelian* quotient singularities.

Warning. Standard usage rarely distinguishes between a variety with algebraic quotient singularities as above and a variety with quotient singularities, by which one usually means that the complete local rings $\hat{\mathcal{O}}_{x,X}$ can be written as rings of invariants of a finite group G acting on a complete regular local ring. (As a further problem, in positive characteristic one could allow finite group schemes acting or one may want to exclude some wild nonlinear actions.) If X has quotient singularities, then X is covered by étale local charts, which are quotients of smooth varieties by finite groups, but it is not true that X has algebraic quotient singularities.

THEOREM 2.23. *Let $g : X \to Y$ be a finite and dominant morphism from a normal variety X to a smooth variety Y, all defined over $\mathbb{C}$. Assume that there is a simple normal crossing divisor $D \subset Y$ such that g is smooth over $Y \setminus D$.*

Then X has algebraic abelian quotient singularities.

Proof. Let us start with a local analytic picture, which is an exact analog of Riemann's method (1.5).

Pick a point $0 \in Y$, and choose an open polydisc $\Delta^n \cong B \subset Y$ centered at 0 with coordinate functions $y_1, \ldots, y_n$ such that $D \cap B \subset (y_1 \cdots y_n = 0)$. Let $B_X \subset g^{-1}(B)$ be a connected component. Set $B^0 := B \setminus (y_1 \cdots y_n = 0)$ and $B_X^0 := B_X \cap g^{-1}(B^0)$.

The fundamental group of B^0 is $\mathbb{Z}^n$. $B_X^0 \to B^0$ is a covering space, corresponding to a finite quotient $\mathbb{Z}^n \to H$. If m is a multiple of $|H|$, then $(m\mathbb{Z})^n \subset \ker[\mathbb{Z}^n \to H]$. The covering of B^0 corresponding to $(m\mathbb{Z})^n$ can be written down explicitly. It is

$$\pi_m : \Delta^n_{v_1, \ldots, v_n} \to B, \quad \text{given by } \pi_m^* y_i = v_i^m.$$

By the Riemann extension theorem we get a factorization

$$
\begin{array}{ccc}
\Delta^n_{v_1, \ldots, v_n} & \xrightarrow{\sigma} & B_X \\
\cong \downarrow & & \downarrow g \\
\Delta^n_{v_1, \ldots, v_n} & \xrightarrow{\pi_m} & B.
\end{array}
$$

Thus B_X is the quotient of Δ^n by the abelian group $\ker[(\mathbb{Z}/m\mathbb{Z})^n \to H]$. So we conclude that X has analytic quotient singularities.

It is not hard to globalize the construction. Pick a point $0 \in Y$. One can choose local coordinates $y_1, \ldots, y_n$ and an affine neighborhood $0 \in U \subset Y$ such that $D \cap U \subset (y_1 \cdots y_n = 0)$ and $(y_1, \ldots, y_n) : U \to \mathbb{C}^n$ is smooth. Set $U_X := g^{-1}(U) \subset X$.

Let $d = \deg g$. The various local degrees $B_X \to B$ are $\leq d$; thus choose $m = d!$ (or any multiple of it). The replacement for the local covers $\Delta^n_{v_1, \ldots, v_n}$ is

$$V := (y_1 - v_1^m = \cdots = y_n - v_n^m = 0) \subset U \times \mathbb{A}^n_{v_1, \ldots, v_n}.$$

Note that V is smooth and V is the normalization of U in the field extension

$$k(V) = k(U, y_1^{1/m}, \ldots, y_n^{1/m})/k(U),$$

which is an abelian Galois extension.

Let $K = k(V) + k(X)$ be the composite of the two field extensions of $k(U)$. Then $K/k(X)$ is Galois with an abelian Galois group. Let Z be the normalization of U in K; thus Z is also the normalization of V in K. We have a basic diagram

$$
\begin{array}{ccc}
Z & \xrightarrow{\sigma} & U_X \\
h \downarrow & & \downarrow g \\
V & \xrightarrow{\pi} & U,
\end{array}
$$

where all maps are finite, dominant and Galois.

It is a nontrivial theorem that if an algebraic variety over $\mathbb{C}$ is normal, then the corresponding analytic space is also normal. This is a special case of the so-called GAGA-type results (after the acronym of the title of [Ser56]), which compare properties of a complex algebraic variety and of the corresponding analytic space.

Once we know that Z is normal as an analytic space, the local computation done at the beginning shows that $h : Z \to V$ has an analytic inverse in an analytic neighborhood of every point. Thus Z is smooth, and so U_X has algebraic quotient singularities. $\qquad\square$

ASIDE 2.24 (Varieties finitely dominated by smooth varieties). There is an interesting open problem related to quotient singularities.

Conjecture 2.24.1. Let $f : X \to Y$ be a finite and dominant morphism from a smooth variety X to a normal variety Y. Then Y has quotient singularities.

This is known and relatively easy in dimension 2 [Bri68] but open already in dimension 3. Many partial results are in [Gur03].

One reason why this is hard is that f itself need not be a quotient map, not even locally. For instance, there are many complicated finite and dominant morphisms $\mathbb{A}^2 \to \mathbb{A}^2$, which are not invariant under any group action.

By thinking of a smooth point as a cone over $\mathbb{P}^{n-1}$, one is led to a projective version of the conjecture, which is known.

Theorem 2.24.2 [Laz84, CMSB02] Let $f : \mathbb{CP}^n \to Y$ be a finite and dominant morphism to a normal variety Y. Then Y is isomorphic to a quotient $\mathbb{P}^n/G$ for some finite subgroup $G \subset PGL(n+1, \mathbb{C})$.

I emphasize again that f itself need not be a quotient map.

2.4. Cyclic quotient singularities

2.25 (Quotients by linear actions). Let G be a finite group acting on an affine variety X and $Y = X/G$. Let $y \in Y$ be a point and $x_1, \ldots, x_m$ its preimages in X. Let $G_i \subset G$ be the stabilizer of x_i. The subgroups G_i are conjugate to each other and

$$
\hat{\mathcal{O}}_{y,Y} = \hat{\mathcal{O}}_{x_i,X}^{G_i}.
$$

Assume now that we are over an algebraically closed field k, $\operatorname{char} k \nmid |G_i|$ and $x_i \in X$ is a closed point.

Let (R, m) be a complete local $k = R/m$-algebra with a G-action. If $\operatorname{char} k \nmid |G|$, the G-action on m is completely reducible, and thus $m \to m/m^2$ has a G-equivariant inverse. This gives a G-equivariant map from the polynomial ring $k[m/m^2]$ to R.

Claim 2.25.1. If R is regular, then it is G-equivariantly isomorphic to the completion of $k[m/m^2]$. Furthermore, R^G is the completion of $k[m/m^2]^G$.

Proof. $k[m/m^2] \to R$ gives surjections $S^i(m/m^2) \twoheadrightarrow m^i/m^{i+1}$. If R is regular, then the two sides have the same dimension, so the maps are isomorphisms. A power series is G-invariant iff each of its homogeneous parts are G-invariant. Thus we get that taking invariants commutes with completion.   □

By (2.18), this implies the following.

Claim 2.25.2. Let k be an algebraically closed field of characteristic zero. Then resolving quotient singularities of algebraic surfaces over k is equivalent to resolving quotients $\mathbb{A}^2/G$ by linear G-actions.   □

Let k be a field and $G \subset GL(n, k)$ a finite subgroup. Then G acts on the affine space $\mathbb{A}^n_k$ and on the polynomial ring $k[x_1, \ldots, x_n]$ by linear transformations.

There is the usual problem that two actions come naturally. If $g = (g_{ij}) \in G$, then one action is matrix multiplication

$$(x_1, \ldots, x_n)^t \mapsto (g_{ij}) \cdot (x_1, \ldots, x_n)^t.$$

On the other hand, if we think of $k[x_1, \ldots, x_n]$ as functions on $\mathbb{A}^n_k$, then the natural action is pulling back by the G-action. This corresponds to

$$(x_1, \ldots, x_n)^t \mapsto (g_{ij})^{-1} \cdot (x_1, \ldots, x_n)^t,$$

which is not an action of G but of the opposite group where we have a new multiplication rule $g \circ h := hg$.

Since here we are interested in abelian group actions, this distinction is rather pointless and we ignore it.

2.26 (Classification of abelian quotients). Let k be an algebraically closed field and $H \subset GL(2, k)$ a finite abelian group, $\operatorname{char} k \nmid |H|$. Our aim is to classify quotient singularities of the form $\mathbb{A}^2/H$.

After a change of coordinates we can assume that H is diagonal; that is, every element is of the form

$$\operatorname{diag}(\epsilon, \eta) := \begin{pmatrix} \epsilon & 0 \\ 0 & \eta \end{pmatrix}, \quad \text{where } \epsilon, \eta \text{ are roots of unity.}$$

Equivalently, $H \subset \mu_n + \mu_n$, where $\mu_n \subset k^*$ denotes the group of nth roots of unity and $n = |H|$.

Assume that $\mathrm{diag}(\epsilon, 1) \in H$, where ϵ is an mth root of unity. Then

$$\mathbb{A}^2_{x,y}/\langle \mathrm{diag}(\epsilon, 1)\rangle \cong \mathbb{A}^2_{x^m,y},$$

and so we can represent $\mathbb{A}^2/H$ as a quotient by a smaller group

$$\mathbb{A}^2_{x,y}/H \cong \mathbb{A}^2_{x^m,y}/\big(H/\langle \mathrm{diag}(\epsilon, 1)\rangle\big).$$

Repeating this procedure, we eventually obtain the following.

Claim 2.26.1. Every abelian quotient singularity $\mathbb{A}^2/G$ is isomorphic to some $\mathbb{A}^2/H$, where for every $1 \neq h \in H$ both eigenvalues of h are different from 1. □

This puts strong restrictions on H. First, for any prime q, the only diagonal subgroup of $GL(2, k)$ isomorphic to $\mathbb{Z}/q + \mathbb{Z}/q$ is $\mu_q + \mu_q$, which contains many elements with 1 as an eigenvalue. Thus H does not contain any subgroup of the form $\mathbb{Z}/q + \mathbb{Z}/q$, and hence H is cyclic. Choose a generator $h \in H$ and a primitive nth root of unity ϵ, where $n = |H|$. Then $h = \mathrm{diag}(\epsilon^a, \epsilon^b)$ for some $a, b \in \mathbb{Z}/n$. As a shorthand, we use $\frac{1}{n}(a, b)$ to denote this group. The corresponding quotient is denoted by

$$\mathbb{A}^2/\tfrac{1}{n}(a, b) := \mathbb{A}^2/\langle \mathrm{diag}(\epsilon^a, \epsilon^b)\rangle.$$

(Our notation does not keep track of the choice of ϵ. We can, however, describe the group $\frac{1}{n}(a, b)$ more invariantly as

$$\tfrac{1}{n}(a, b) = \mathrm{im}[\mu_n \to \mu_n + \mu_n : \eta \mapsto (\eta^a, \eta^b)].$$

Thus it is independent of the choices that we made.)

If $(n, a) \neq 1$, then $h^{n/(n,a)}$ contradicts (2.26.1) and similarly if $(n, b) \neq 1$. Finally, if $(n, a) = 1$, then we can solve $ac \equiv 1 \mod n$ and $\frac{1}{n}(a, b) = \frac{1}{n}(1, bc)$. Thus we proved the existence part of the final result.

THEOREM 2.27. *Let k be an algebraically closed field and $H \subset GL(2, k)$ a finite abelian group, char $p \nmid |H|$. Then*

(1) $\mathbb{A}^2/H$ *is isomorphic to a quotient of the form $\mathbb{A}^2/\frac{1}{n}(1, q)$ where $(n, q) = 1$, and*

(2) $\mathbb{A}^2/\frac{1}{n}(1, q) \cong \mathbb{A}^2/\frac{1}{n}(1, q')$ *iff $q \equiv q' \mod n$ or $qq' \equiv 1 \mod n$.*

Proof. We already saw the first part.

For the second part we need a geometric way to recover the group $\frac{1}{n}(1, q)$ from the quotient $\mathbb{A}^2/\frac{1}{n}(1, q)$. The two possibilities for q' then come from

$$\tfrac{1}{n}(1, q) = \tfrac{1}{n}(q', qq') = \tfrac{1}{n}(q', 1).$$

Assume for simplicity that we are over $\mathbb{C}$. Since $\mathbb{C}^2 \setminus (0, 0)$ is connected and simply connected, we conclude that $\mathbb{C}^2 \setminus (0, 0)$ is the universal cover of $(\mathbb{C}^2 \setminus (0,0))/\frac{1}{n}(1,q)$. This gives an intrinsic way to recover $\mathbb{C}^2$ and the group action from the quotient.

The same idea works in general once we work out the relevant notion of fundamental group. One has to be careful, though, since in positive characteristic $\mathbb{A}^2$ itself is not simply connected, but its completion at any point $\hat{\mathbb{A}}^2$ is. $\square$

ASIDE 2.28 (Classification of all quotient singularities). For a good general introduction, see [Ben93] or [Smi95].

Let $G \subset GL(n,k)$ be a finite group, again assuming char $k \nmid |G|$. An element $1 \neq g \in GL(n,k)$ is called a *quasi-reflection* if all but one of its eigenvalues are 1.

Theorem 2.28.1. [ST54] Let $G \subset GL(n,k)$ be a finite subgroup generated by quasi-reflections. Then $\mathbb{A}^n/G \cong \mathbb{A}^n$.

For any $G \subset GL(n,k)$, the subgroup $G_{qr} \subset G$ generated by quasi-reflections is normal, and so
$$\mathbb{A}^n/G = (\mathbb{A}^n/G_{qr})/(G/G_{qr}) \cong \mathbb{A}^n/(G/G_{qr}).$$

Thus it is enough to classify quotients by subgroups containing no quasi-reflections. As in (2.27) we obtain the following.

Theorem 2.28.2. Let k be an algebraically closed field and $G_1, G_2 \subset GL(n,k)$ two subgroups containing no quasi-reflections. If char $k \nmid |G_i|$, then the following are equivalent:

(1) $\mathbb{A}^n/G_1 \cong \mathbb{A}^n/G_2$,
(2) $\hat{\mathbb{A}}^n/G_1 \cong \hat{\mathbb{A}}^n/G_2$, and
(3) G_1 and G_2 are conjugate in $GL(n,k)$. $\square$

2.29 (How not to resolve quotient singularities). The most straightforward idea to resolve a quotient singularity $\mathbb{A}^2_{x,y}/\frac{1}{n}(a,b)$ is to blow up $\mathbb{A}^2$ at the origin and see what happens. The group action lifts to the blow-up, and we get two affine charts
$$\mathbb{A}^2_{x,\frac{y}{x}}/\tfrac{1}{n}(a, b-a) \quad \text{and} \quad \mathbb{A}^2_{\frac{x}{y},y}/\tfrac{1}{n}(a-b, b).$$

We represent this step symbolically as
$$\tfrac{1}{n}(a,b) \leftarrow \left(\tfrac{1}{n}(a, b-a), \tfrac{1}{n}(a-b, b)\right).$$

It looks like we may get a Euclidean algorithm on the pair (a,b). In some cases, we get a resolution right away, as in
$$\tfrac{1}{n}(a,a) \leftarrow \left(\tfrac{1}{n}(a,0), \tfrac{1}{n}(0,a)\right).$$

Furthermore, $\mathbb{Z}/2, \mathbb{Z}/3$ and $\mathbb{Z}/4$ quotients are all resolved this way. For instance, starting with $\frac{1}{3}(1,2)$ we get
$$\tfrac{1}{3}(1,2) \leftarrow \left(\tfrac{1}{3}(1,1), \tfrac{1}{3}(2,2)\right),$$

and both of the latter are resolved with one more blow-up. Similarly we
have
$$\tfrac{1}{4}(1,3) \leftarrow \left(\tfrac{1}{4}(1,2) \cong \tfrac{1}{2}(1,1), \tfrac{1}{4}(2,3) \cong \tfrac{1}{2}(1,1)\right),$$
and both of the latter are resolved with one more blow-up. More generally,
we have the following.

Claim 2.29.1. Any $\mathbb{Z}/(2^m 3^n)$-quotient is resolved this way.

Proof. By (2.27) we may assume that the singularity is of the form
$\mathbb{A}^2/\frac{1}{2^m 3^n}(1,b)$ with $(2^m 3^n, b) = 1$. If $m \geq 1$, then b is odd and the first
blow-up is
$$\mathbb{A}^2/\tfrac{1}{2^m 3^n}(1,b) \leftarrow \left(\mathbb{A}^2/\tfrac{1}{2^m 3^n}(1, b-1), \mathbb{A}^2/\tfrac{1}{2^m 3^n}(1-b, b)\right).$$
Since $b - 1$ is even, using (2.26.1) we have $\mathbb{Z}/(2^{m-1}3^n)$-quotients on the
right-hand side.

Similarly we see that after two blow-ups we can reduce 3^n to 3^{n-1}. □

The above process, however, breaks down for $\mathbb{Z}/5$ quotients. Consider
the blow-up
$$\tfrac{1}{5}(1,2) \leftarrow \left(\tfrac{1}{5}(1,1), \tfrac{1}{5}(4,2)\right).$$
The first of the new singularities is resolved by one more blow-up, but the
second one is
$$\tfrac{1}{5}(4,2) = \tfrac{1}{5}(3 \cdot 4, 3 \cdot 2) = \tfrac{1}{5}(2,1) \cong \tfrac{1}{5}(1,2).$$
Thus we get back our original singularity. More generally, this happens for
every $p \geq 5$.

Claim 2.29.2. Let $p \geq 5$ be a prime. The only $\mathbb{Z}/p$-quotient resolved as
above is $\frac{1}{p}(1,1)$.

Proof. Starting with $\frac{1}{p}(1, p-1)$, we get a sequence of blow-ups, where
we write only one of the charts:
$$\tfrac{1}{p}(1, p-1) \leftarrow \tfrac{1}{p}(1, p-2) \leftarrow \cdots \leftarrow \tfrac{1}{p}(1,2).$$
We have run through all $\mathbb{Z}/p$-quotients save $\frac{1}{p}(1,1)$. At the next blow-up
we get
$$\tfrac{1}{p}(1,2) \leftarrow \left(\tfrac{1}{p}(1,1), \tfrac{1}{p}(p-1,2) \cong \tfrac{1}{p}(1, p-2)\right).$$
We are back in the first sequence and from now on go through the loop
forever. □

EXAMPLE 2.30 (Failure of simultaneous resolution). Let $p \geq 5$ be a
prime and $2 \leq a < p$. There is no commutative diagram

$$\begin{array}{ccc}
X & \xrightarrow{f} & \mathbb{C}^2 \\
h \downarrow & & \downarrow \\
Y & \xrightarrow{g} & \mathbb{C}^2/\tfrac{1}{p}(1,a),
\end{array}$$

where f, g are proper and birational, h is finite and X, Y are both smooth.

Indeed, $k(X) = k(\mathbb{C}^2)$ and $k(Y) = k(\mathbb{C}^2/\frac{1}{p}(1, a))$. Thus $k(X)/k(Y)$ is a Galois extension, and the Galois action descends to a $\mathbb{Z}/p$-action on X since X is the normalization of Y in $k(X)$. By (2.13) X is obtained from $\mathbb{C}^2$ by a sequence of blow-ups, but we just computed that we can never get rid of singularities of the form $\mathbb{C}^2/\frac{1}{p}(1, b)$ on the quotient by blowing up points.

2.31 (Resolving quotient singularities). $\mathbb{A}^2_{x,y}/\frac{1}{n}(1, q)$ is smooth only for $q = 0$, that is, if the action is trivial on one of the coordinates. With this in mind, it may be helpful to blow up $\mathbb{A}^2$ in such a way that we get a trivial action on the exceptional curve. The simplest blow-up with this property is $B_{(x^q, y)}\mathbb{A}^2$. We can define this blow-up as

$$B_{(x^q, y)}\mathbb{A}^2 = (sx^q = ty) \subset \mathbb{A}^2_{x,y} \times \mathbb{P}^1_{s,t}$$

and take the trivial group action on $\mathbb{P}^1$. It is covered by two affine charts. One of them has coordinates $x, y, w = s/t$, and here the equation becomes $wx^q = y$. Thus we can eliminate y and get a smooth quotient

$$\mathbb{A}^3_{x,y,w}/\tfrac{1}{n}(1, q, 0) \supset (wx^q = y)/\tfrac{1}{n}(1, q, 0) \cong \mathbb{A}^2_{y,w}/\tfrac{1}{n}(q, 0) \cong \mathbb{A}^2_{y^n,w}.$$

The other chart has coordinates $x, y, z = t/s$, and we get the quotient

$$\mathbb{A}^3_{x,y,z}/\tfrac{1}{n}(1, q, 0) \supset (x^q = yz)/\tfrac{1}{n}(1, q, 0).$$

Even before taking the quotient we have a singular surface with equation $x^q = yz$. This itself can be written as a quotient

$$(x^q = yz) \cong \mathbb{A}^2_{u,v}/\tfrac{1}{q}(1, -1), \quad \text{where } x = uv, y = v^q, z = u^q.$$

By luck, the $\frac{1}{n}(1, q, 0)$-action on $(x^q = yz)$ lifts as the $\frac{1}{n}(0, 1)$-action on $\mathbb{A}^2_{u,v}$. Thus

$$(x^q = yz)/\tfrac{1}{n}(1, q, 0) \cong \mathbb{A}^2_{u,v}/\left(\tfrac{1}{q}(1, -1) \times \tfrac{1}{n}(0, 1)\right).$$

We can compute the right-hand side by first taking the quotient by the $\frac{1}{n}(0, 1)$-action, giving

$$\mathbb{A}^2_{u,v}/\tfrac{1}{n}(0, 1) \cong \mathbb{A}^2_{u,v^n}.$$

Thus

$$(x^q = yz)/\tfrac{1}{n}(1, q, 0) \cong \mathbb{A}^2_{u,v}/\tfrac{1}{q}(1, -n).$$

In order to bring this into our normal form, write

$$-n = -b_1 q + r, \quad \text{or, equivalently} \quad \tfrac{n}{q} = b_1 - \tfrac{r}{q},$$

where $0 < r < q$. Thus we obtain the following.

PROPOSITION 2.32. *The partial resolution* $X_1 \to \mathbb{A}^2/\frac{1}{n}(1,q)$ *obtained as the quotient*

$$X_1 := (B_{(x^q,y)}\mathbb{A}^2)/\tfrac{1}{n}(1,q) \to \mathbb{A}^2/\tfrac{1}{n}(1,q)$$

has a unique singular point isomorphic to

$$\mathbb{A}^2/\tfrac{1}{q}(1,r), \quad \text{where } 0 < r < q \text{ and } \tfrac{n}{q} = b_1 - \tfrac{r}{q}.$$

Iterating this procedure gives a resolution of the singularity $\mathbb{A}^2/\frac{1}{n}(1,q)$. ☐

The above resolution process is thus governed by the modified continued fraction expansion

$$\frac{n}{q} = b_0 - \cfrac{1}{b_1 - \cfrac{1}{b_2 - \cfrac{1}{b_3 - \cfrac{1}{\cdots}}}}$$

Let $b_0, \ldots, b_s$ be the entries in the above continued fraction. It is not hard to compute that one gets the minimal resolution, there are $s+1$ exceptional curves $E_0, \ldots, E_s$, and the nonzero intersection numbers between them are $(E_i \cdot E_i) = -b_i$ for $i = 0, \ldots, s$, and $(E_i \cdot E_{i+1}) = 1$ for $i = 0, \ldots, s-1$. Thus the E_i form a chain of rational curves with dual graph:

$$b_0 \; - \; b_1 \; - \; \cdots \; - \; b_{s-1} \; - \; b_s.$$

Conversely, one can see that every normal surface singularity whose dual graph is a chain of rational curves is a cyclic quotient singularity, see [Bri68].

2.5. The Albanese method using projections

The higher-dimensional version of the method of Albanese (Section 1.3) proves that every projective variety X is birational to another variety X', where every point of X' has multiplicity $\leq (\dim X)!$. For surfaces this was proved in [Alb24b]. A modern exposition was given in a 1963 lecture by Artin and later generalized to higher dimensions in [Abh66, Sec.12]. Our treatment is close to [Lip75, p.200].

If $\dim X = 1$, then we get that X' is smooth. If $\dim X = 2$, then X' has only double points, and we see in Section 2.6 that these are quite easy to resolve, at least in all characteristics different from 2.

If $\dim X = 3$, then X' has points of multiplicity at most 6. Points of multiplicity at most 6 are no doubt rather special among all singularities, but they do not seem to be any easier to resolve in characteristic zero than any other singularity. In positive characteristic, one gets some advantage, though. Points of multiplicity m can, at least after completion, be realized as sitting on a degree m cover of $\mathbb{A}^n$. If m is less than the characteristic, then the results of Section 1.12 show that we avoid complications due to

wild ramification. The method of maximal contact also works when the
multiplicity is less than the characteristic; see (2.57).

This enabled Abhyankar [Abh66] to prove resolution of 3-dimensional
varieties in characteristic > 3!.

We start with the geometric version of the algorithm. The abstract case
is discussed in (2.44). The notion of multiplicity and its basic properties
are reviewed in Section 2.8.

ALGORITHM 2.33 (Albanese algorithm). Let $X_0 \subset \mathbb{P}^N$ be an irreducible
and reduced projective variety of dimension n. Assume that $X_i \subset \mathbb{P}^{N-i}$
and $\tau_i : X_0 \dashrightarrow X_i$ are already defined. Pick any point $p_i \in X_i \subset \mathbb{P}^{n-i}$
such that

$$\deg(X_0/X_i) \cdot \operatorname{mult}_{p_i} X_i \geq n! + 1,$$

and let $\pi_i : \mathbb{P}^{N-i} \dashrightarrow \mathbb{P}^{N-i-1}$ be the projection from the point p_i. Set

$$X_{i+1} := \overline{\pi_i(X_i)},$$

and let $\tau_{i+1} := \pi_i \circ \tau_i : X_0 \dashrightarrow X_{i+1}$ be the composite.

REMARK 2.34. It seems very surprising that the Albanese algorithm
can be used to improve nonisolated singularities. If we think of a projection
as a variant of blowing up a point, then it is quite hard to imagine how it
could eliminate nonisolated singularities.

There is, however, a slim chance that this might work. Let $Z_0 \subset X_0$
be a positive dimensional component of the singular locus and $Z_i \subset X_i$ its
image in X_i.

With each projection the degree of Z_i drops, and so if we are very
lucky, then eventually Z_i becomes a linear subvariety of X_i. At the next
step, when we project from a point of Z_i, Z_i gets contracted to a lower-
dimensional subvariety. Thus eventually we could eliminate all high-multi-
plicity components of the singular locus.

The final version of the method actually should have a very good chance
of doing just this. Here we start with $X \subset \mathbb{P}^s$ and embed it by the global
sections of $\mathcal{O}_{\mathbb{P}^s}(m)$ for $m \gg 1$.

If $Z \subset X$ is any subvariety, then under such an embedding the image
of Z spans a linear space $L(Z)$ whose dimension grows asymptotically as

$$\frac{\deg Z}{(\dim Z)!} m^{\dim Z}.$$

Thus for $m \gg 1$, the image of the singular locus of X sits inside $L(Z)$,
whose dimension is very small compared to the dimension of the ambient
space. Since we always project from a point of $L(Z)$, we expect that $L(Z)$
eventually gets mapped to a point if there are enough projections to do.

On the other hand, each time we blow up, we usually create new singularities that lie outside the image of $L(Z)$, so the above argument is not a proof of anything.

THEOREM 2.35 (Weak Albanese theorem). *Let $X_0 \subset \mathbb{P}^N$ be an irreducible, reduced projective variety of dimension n spanning $\mathbb{P}^N$ defined over an algebraically closed field.*

If $\deg X_0 < (n!+1)(N+1-n)$, then the Albanese algorithm eventually stops with a variety X_i and a map $\tau_i : X_0 \dashrightarrow X_i$ such that

(1) *either $\deg(X_0/X_i) \cdot \mathrm{mult}_p X_i \leq n!$ for every $p \in X_i$,*
(2) *or X_i is a cone and $\deg(X_0/X_i) \leq n!$.*

COROLLARY 2.36. *Every irreducible, reduced projective surface X over an algebraically closed field is birational to a normal surface X' whose singularities are finitely many double points.*

Proof. As in the curve case (1.12), first we need to get an embedding $X \hookrightarrow \mathbb{P}^N$, where the Albanese algorithm (2.33) can start. This is easy in all dimensions using (2.37).

By (2.35), the sequence of projections eventually stops, and we have four possible outcomes.

(1) $\tau_i : X \dashrightarrow X_i$ is birational, and every point of X_i has multiplicity at most 2. Let X' be the normalization of X_i. It has only finitely many double points by (2.72.2), and it is birational to X.

(2) $\tau_i : X \dashrightarrow X_i$ has degree 2, and every point of X_i has multiplicity 1. Let X' be the normalization of X_i in the quotient field of X. It has only finitely many double points by (2.72.2), and it is birational to X.

(3) $\tau_i : X \dashrightarrow X_i$ is birational, and X_i is a cone over a curve C_i. Let $C_i' \to C_i$ be a resolution of singularities. Then X is birational to the smooth variety $\mathbb{P}^1 \times C_i'$.

(4) $\tau_i : X \dashrightarrow X_i$ has degree 2, and X_i is a cone over a curve C_i. Let X' be the normalization of $\mathbb{P}^1 \times C_i'$ in the quotient field of X. It has only finitely many double points by (2.72.2), and it is birational to X.   □

LEMMA 2.37. *Let X be a projective variety of dimension n and L an ample line bundle on X. Then*

$$h^0(X, L^{\otimes m}) > \frac{(L^n)}{n!} m^n + O(m^{n-1}).$$

Proof. This follows from the Riemann-Roch theorem on singular varieties (see, for instance, [Ful84, Chap.18]). Indeed, the Riemann-Roch theorem states that $\chi(X, L^{\otimes m})$ is a polynomial in m whose leading term is

$\frac{(L^n)}{n!} m^n$. This is all we need since $\chi(X, L^{\otimes m}) = h^0(X, L^{\otimes m})$ for $m \gg 1$ by Serre's vanishing theorem.

One can also use the asymptotic Riemann-Roch theorems for coherent sheaves (see, for instance, [Kol96, VI.2.15]).

It is, however, worthwhile to note that there is a completely elementary argument giving this inequality.

We may assume that L is very ample and embed X into $\mathbb{P}^N$ by L. Then project it generically to a hypersurface of degree (L^n) to get $\pi : X \to X' \subset \mathbb{P}^{n+1}$. For $m \geq (L^n)$,

$$
\begin{aligned}
h^0(X, L^m) & \geq h^0\big(X', \mathcal{O}_{\mathbb{P}^{n+1}}(m)|_{X'}\big) \\
& \geq h^0\big(\mathbb{P}^{n+1}, \mathcal{O}_{\mathbb{P}^{n+1}}(m)\big) - h^0\big(\mathbb{P}^{n+1}, \mathcal{O}_{\mathbb{P}^{n+1}}(m - (L^n))\big) \\
& = \binom{n+m+1}{n+1} - \binom{n+m+1-(L^n)}{n+1}.
\end{aligned}
$$

This can be further estimated from below as

$$
\begin{aligned}
\binom{n+m+1}{n+1} - \binom{n+m+1-(L^n)}{n+1} & = \sum_{j=1}^{(L^n)} \binom{n+m+1-j}{n} \\
& \geq (L^n)\binom{n+m+1-(L^n)}{n} \geq (L^n)\frac{(m+1-(L^n))^n}{n!}. \quad \square
\end{aligned}
$$

ASIDE 2.38 (Effective Riemann-Roch). We do not need it here, but it is sometimes useful to know the following effective version of the Riemann-Roch theorem.

Theorem 2.38.1. [KM83] There is an effectively computable $C(x, y, z)$ such that if X is a normal projective variety of dimension n and L an ample line bundle on X, then

$$
\left| h^0(X, L^{\otimes m}) - \frac{(L^n)}{n!} m^n + \frac{(K_X \cdot L^{n-1})}{2(n-1)!} m^{n-1} \right| \leq C \cdot m^{n-2}
$$

for every $m \geq 1$, where the constant C depends only on n, (L^n), $(K_X \cdot L^{n-1})$.

As far as I know, $C(x, y, z)$ was never actually computed, and it is quite large. In applications the uniform asymptotic behavior seems the most important. There does not even appear to be conjectures about the nature of the optimal bound $C(x, y, z)$.

Before we start the proof of (2.35), we need some elementary lemmas about varieties and their projections.

2.39 (Projections of varieties). Let $X \subset \mathbb{P}^N$ be any irreducible, reduced variety and $p \in X$ a point. Let $\pi : \mathbb{P}^N \dashrightarrow \mathbb{P}^{N-1}$ denote the projection from p and $X_1 := \overline{\pi(X)}$ the closure of the image of X. Assume that X is not a cone with vertex p.

It is straightforward to generalize (1.15) to higher dimensions, and we obtain that

  (1) if X spans $\mathbb{P}^N$, then X_1 spans $\mathbb{P}^{N-1}$,

(2) $\deg X_1 \cdot \deg(X/X_1) = \deg X - \mathrm{mult}_p X$.

LEMMA 2.40 (Bertini). *Let k be an algebraically closed field. Let $X \subset \mathbb{P}_k^N$ be reduced, of pure dimension n, with r irreducible components and not contained in any hyperplane. Then $\deg X \geq N + 1 - rn$.*

Equivalently, if L is a very ample line bundle on X, then $h^0(L) \leq (L^n) + rn$.

Proof. It is enough to prove that $r = 1$ case.

If X is a cone with vertex p, then we can project from p and use induction on $\dim X$.

If X is not a cone with vertex p, then let $\pi : X \dashrightarrow X_1$ be the projection from p. By induction on N and using (2.39) we get that

$$\begin{aligned} \deg X &\geq \deg X_1 \cdot \deg(X/X_1) + \mathrm{mult}_p X \\ &\geq (N - n) \cdot \deg(X/X_1) + 1 \geq N - n + 1. \quad \square \end{aligned}$$

Note also that X being reduced is an important condition. There are double lines in $\mathbb{P}^n$ that are not contained in any hyperplane.

ASIDE 2.41 (Minimal degree varieties). The above argument also shows that if X is a variety of minimal degree, that is, if $\deg X = N + 1 - \dim X$, and X is not a cone, then X is smooth and its projection is also a variety of minimal degree.

Starting with this observation, Del Pezzo in 1886 for surfaces and Bertini in 1907 in general classified all varieties $X \subset \mathbb{P}^n$ such that $\deg X = N + 1 - \dim X$. These are

(1) $\mathbb{P}^m$,
(2) quadric hypersurfaces $Q \subset \mathbb{P}^m$,
(3) the Veronese surface $\mathbb{P}^2 \subset \mathbb{P}^5$, embedded by $|\mathcal{O}_{\mathbb{P}^2}(2)|$,
(4) scrolls, $\mathrm{Proj}_{\mathbb{P}^1}(\sum \mathcal{O}_{\mathbb{P}^1}(a_i))$ for $a_i > 0$, embedded by their $|\mathcal{O}(1)|$,
(5) cones over one of the examples in (3) or (4).

We recommend the survey paper [EH87] for a leisurely introduction to their study.

It is harder to show that the inequality $h^0(L) \leq (L^n) + n$ holds if L is ample (but not very ample) and to prove that if $h^0(L) = (L^n) + n$, then L is very ample, and so we have a variety on the Del Pezzo-Bertini list.

It is also possible to classify those pairs (X, L) for which $h^0(L) = (L^n) + n - 1$ or $h^0(L) = (L^n) + n - 2$. These lists are longer, but they contain many geometrically interesting varieties. The lecture notes [Fuj90] provide an introduction to this and related topics.

2.42 (Proof of (2.35)). Starting with $X_0 \subset \mathbb{P}^N$ such that $\deg X_0 < (n! + 1)(N + 1 - n)$, we get a sequence of varieties $X_i \subset \mathbb{P}^{N-i}$.

We prove by induction on i that

$$\deg(X_0/X_i) \cdot \deg X_i < (n! + 1)(N - i + 1 - n). \qquad (2.42.1)$$

In order to get X_{i+1}, pick a point $p_i \in X_i$ such that

$$\deg(X_0/X_i) \cdot \text{mult}_{p_i} X_i \geq n! + 1,$$

and let $\pi_i : \mathbb{P}^{N-i} \dashrightarrow \mathbb{P}^{N-i-1}$ be the projection from the point p_i.

If X_i is not a cone with vertex p_i, then $\pi_i : X_i \dashrightarrow X_{i+1}$ is generically finite and

$$\deg X_{i+1} \cdot \deg(X_i/X_{i+1}) = \deg X_i - \text{mult}_{p_i} X_i$$

by (2.39). Thus

$$
\begin{aligned}
\deg(X_0/X_{i+1}) \cdot \deg X_{i+1} &= \\
&= \deg(X_0/X_i) \cdot \deg(X_i/X_{i+1}) \cdot \deg X_{i+1} \\
&\leq \deg(X_0/X_i) \cdot \deg X_i - \deg(X_0/X_i) \, \text{mult}_{p_i} X_i \\
&\leq \deg(X_0/X_i) \cdot \deg X_i - (n! + 1) \\
&\leq (n! + 1)(N - i + 1 - n) - (n! + 1) \\
&= (n! + 1)\big(N - (i + 1) + 1 - n\big).
\end{aligned}
$$

The sequence of projections eventually must stop, so we either stop with X_i such that $\deg(X_0/X_i) \cdot \text{mult}_p X_i \leq n!$ for every $p \in X_i$ or we stop when X_i becomes a cone.

Since X_i spans $\mathbb{P}^{N-i}$, $\deg X_i \geq N - i + 1 - n$ by (2.40), and together with (2.42.1) this shows that $\deg(X_0/X_i) \leq n!$, as needed. $\qquad\square$

2.43 (Abstract projection). Mostly for the experience rather than for any intrinsic reason, let us rewrite the Albanese algorithm (2.33) in abstract terms. Instead of working with projections $X_i \dashrightarrow X_{i+1}$ that are not birational, we want to stay with birationally equivalent varieties all the time.

Fix a field k. We work with pairs (X, L), where X is a normal k-variety and L is ample and generated by global sections. Later we also need to assume that $X_{\bar{k}}$ is reduced, which always holds as long as k is perfect. What is the analog of a projection?

Given a point $p \in X$, we want to create a pair (X', L') such that $H^0(X', L')$ is naturally isomorphic to the subspace of sections of $H^0(X, L)$ that vanish at p.

If L is very ample and p is a k-point, the first approximation is given by $B_p X$. In general, if we look at all the sections of $H^0(X, L)$ that vanish at p, they do not generate the maximal ideal at p and they may vanish at other points as well. Thus the natural thing to do is to blow up the subsheaf of L generated by these sections, that is,

$$J := \text{im}[H^0(X, L \otimes m_p) \otimes \mathcal{O}_X \to L].$$

Let $B_J X := \text{Proj}_X \sum_{s \geq 0} J^s$ be the blow-up of J with projection $\pi : B_J X \to X$. Essentially by definition, $M := \pi^{-1} J \subset \pi^* L$ is locally free, generated by global sections and

$$h^0(B_J X, M) = h^0(X, L \otimes m_p) \geq h^0(X, L) - \deg p,$$

where $\deg p = \dim_k k(p)$ is the degree of the point p. We can compute the self-intersection (L^n) by using n general sections in $H^0(X, L \otimes m_p)$. The "same" sections can be used to compute the self-intersection (M^n), but we lose all the local contributions coming from the support of L/J. This gives that

$$(M^n) = (L^n) - \mathrm{mult}_J X,$$

where $\mathrm{mult}_J X$ is defined in (2.69). The only problem is that M may not be ample on $B_J X$. By Stein factorization, there is a unique $f : B_J X \to X'$ such that $f_* \mathcal{O}_{B_J X} = \mathcal{O}_{X'}$ and $M = f^* L'$ for some ample and globally generated line bundle L' on X'.

Then $(X, L) \dashrightarrow (X', L')$ is the required analog of the projection.

ALGORITHM 2.44 (Abstract Albanese algorithm). Let k be a field and $c_1, c_2 > 0$ constants. Let X be an irreducible and reduced projective k-variety of dimension n such that $X_{\bar{k}}$ is reduced and has r irreducible components. Let L be an ample and globally generated line bundle on X such that

$$h^0(X, L) \geq c_1^{-1}(L^n) + rn.$$

Set $X_0 := X, L_0 := L$. Assume that we have already defined an irreducible and reduced projective variety X_i and an ample and globally generated line bundle L_i on X_i.

Pick any point $p_i \in X_i$ such that $\mathrm{mult}_{p_i} X_i \geq c_2 \deg p_i$. Following (2.43), set

$$J_i := \mathrm{im}[H^0(X_i, L_i \otimes m_{p_i}) \otimes \mathcal{O}_{X_i} \to L_i],$$

and let $Y_i := \mathrm{Proj}_{X_i} \sum_{s \geq 0} J_i^s$ be the blow-up of J_i with projection $\pi_i : Y_i \to X_i$.

As in (2.43), $M_i := \pi_i^{-1} J_i$ is locally free, generated by global sections,

$$h^0(Y_i, \pi_i^{-1} J_i) = h^0(X_i, L_i) - \deg p_i$$

and

$$(M_i^n) = (L_i^n) - \mathrm{mult}_{J_i} X_i \leq (L_i^n) - c_2 \deg p_i.$$

Finally let $Y_i \to X_{i+1}$ be the Stein factorization of the map given by M_i and L_{i+1} the corresponding line bundle on X_{i+1}.

THEOREM 2.45 (Strong Albanese theorem). *Let the notation be as above. Assume that*

(1) $h^0(X, L) \geq c_1^{-1}(L^n) + rn$ *and*

(2) $(C \cdot L) > c_1 c_2/(c_2 - c_1)$ *for every curve* $C \subset X_{\bar{k}}$.

Then the Albanese algorithm (2.44) eventually stops with a variety X_j *such that* $\mathrm{mult}_p X_j < c_2 \deg p$ *for every* $p \in X_j$.

Proof. The algorithm constructs a sequence of pairs (X_i, L_i), and $h^0(X_i, L_i)$ drops at each step. Thus eventually the procedure stops, because

(3) either $\text{mult}_p X_j < c_2 \deg p$ for every $p \in X_j$ (and so we are done),

(4) or $\dim X_{j+1} < \dim X_j = \dim X$.

We only have to exclude the latter possibility.

Set $D := \sum_{i=0}^j \deg p_i$. By construction

$$h^0(X_{j+1}, L_{j+1}) \geq h^0(X_0, L_0) - D \quad \text{and} \quad 0 = (L_{j+1}^n) \leq (L_0^n) - c_2 D.$$

The latter implies that $D \leq c_2^{-1}(L_0^n)$. Thus

$$\begin{aligned}
h^0(X_{j+1}, L_{j+1}) &= h^0(X_0, L_0) - D \\
&\geq c_1^{-1}(L_0^n) + rn - c_2^{-1}(L_0^n) = (c_1^{-1} - c_2^{-1})(L_0^n) + rn.
\end{aligned}$$

On the other hand, if $\dim X_{j+1} = m < n$, then by (2.40),

$$(L_{j+1}^m) \geq h^0(X_{j+1}, L_{j+1}) + 1 - rm.$$

This means that, over $\bar{k}$, an intersection of m general members $H_1, \ldots, H_m \in |L_{j+1}|$ consists of at least

$$(L_{i+1}^m) \geq h^0(X_{i+1}, L_{i+1}) - rm \geq (c_1^{-1} - c_2^{-1})(L_0^n)$$

distinct m-dimensional subvarieties $\{F_s : s \in S\}$. The birational transforms of H_j are $H_1', \ldots, H_m' \in |L_0|$, and their intersection consists of the birational transforms F_s' of F_s plus an unknown subscheme along which $X_0 \dashrightarrow X_i$ is not a local isomorphism. Let $C_s \subset F_s'$ be the intersection of $m-1$ members of $|L_0|$ restricted to F_s. By (2.46) we get that

$$(L_0^n) \geq \sum_{s \in S}(C_s' \cdot L_0) \geq |S| \cdot \min_{C \subset X_0}(C \cdot L_0) > (c_1^{-1} - c_2^{-1})(L_0^n) \cdot \frac{c_1 c_2}{c_2 - c_1} = (L_0^n).$$

This contradiction proves that (2.45.4) is impossible. $\qquad\square$

LEMMA 2.46. Let X be an n-dimensional variety and L a nef line bundle on X. Let $g_1, \ldots, g_{n-1}$ be global sections of L. Let $\{C_s : s \in S\}$ be the 1-dimensional irreducible components of $(g_1 = \cdots = g_{n-1} = 0)$. Then

$$\sum_{s \in S}(C_s \cdot L) \leq (L^n).$$

Proof. This is clear for $n = 1$. Next we use induction on $n \geq 2$. If all the $g_i = 0$, then $S = \emptyset$ and we are done since $0 \leq (L^n)$. So we may assume that $g_{n-1} \neq 0$ and let $Y_j : j \in J$ be the irreducible components of $(g_{n-1} = 0)$. Then, by induction,

$$\sum_{s \in S}(C_s \cdot L) \leq \sum_j \left(\sum_{C_s \subset Y_j}(C_s \cdot L) \right) \leq \sum_j (Y_j \cdot L^{n-1}) = (L^n). \qquad \square$$

COROLLARY 2.47. *Every irreducible, reduced projective variety X over a perfect field k is birational to a normal variety X' such that*

$$\operatorname{mult}_p X' \leq (\dim X)! \quad \text{for every } p \in X'(\bar{k}).$$

Proof. Set $X_0 = X$, and let L be any ample line bundle on X. By (2.37), we can take $c_1 = n! + \epsilon$, and then (2.45.1) holds for $L_0 := L^{\otimes m}$ for every $m \gg 1$. Take $c_2 = n! + 1$.

Note that every curve in X_0 has degree at least m with respect to $L^{\otimes m}$; thus if $m > c_1 c_2/(c_2 - c_1) = (n! + \epsilon)(n! + 1)(1 - \epsilon)^{-1}$, then (2.45.2) is also satisfied.

Thus we get a normal variety X' such that $\operatorname{mult}_p X' \leq (\dim X)!$ for every $p \in X'(\bar{k})$ by (2.45). $\square$

2.6. Resolving double points, char $\neq 2$

Let S be an algebraic surface over an algebraically closed field k, and assume that S has only points of multiplicity 1 and 2. By (2.72.3), its normalization $\bar{S} \to S$ also has only points of multiplicity ≤ 2, and it has only finitely many singular points $p_1, \ldots, p_n \in \bar{S}$. By (2.18) we need to construct resolutions of the completions of S at the points p_i. We start with the following result describing double points of normal varieties.

PROPOSITION 2.48. *Let k be an algebraically closed field, X a normal variety of dimension n and $x \in X$ a point of multiplicity m. Then its complete local ring can be written as*

$$\hat{\mathcal{O}}_{x,X} \cong \text{normalization of } k[[x_1, \ldots, x_n, y]]/f(x_1, \ldots, x_n, y),$$

where $f(\mathbf{x}, y) = y^m + r_1(\mathbf{x})y^{m-1} + \cdots + r_m(\mathbf{x})$ and $r_i(\mathbf{x}) \in k[[x_1, \ldots, x_n]]$.

Proof. The question is local so we may assume that X is affine. Let $\pi : X \to \mathbb{A}^{n+1}$ be a general projection. The image $\pi(X)$ is a hypersurface, and the projection $\pi : X \to \pi(X)$ is finite and birational. By (2.77), $\pi(X)$ has multiplicity m at $\pi(x)$, which we choose as the origin.

The equation of $\pi(X)$ contains a monomial of degree m, and after a general coordinate change we may assume that y^m appears with nonzero coefficient. Thus the Weierstrass preparation theorem (1.93) shows that after completion $\pi(X)$ is given by an equation

$$y^m + r_1(\mathbf{x})y^{m-1} + \cdots + r_m(\mathbf{x}) = 0.$$

We still need the result that the complete local ring of a normal point is again normal (cf. [Mat89, 32.2]) and so the completion of $\mathcal{O}_{x,X}$ is the same as the normalization of the completion of $\mathcal{O}_{\pi(x),\pi(X)}$. $\square$

While this is a very useful result, in practice it may be very difficult to compute the normalization of a ring, except for double points.

PROPOSITION 2.49. *Let k be an algebraically closed field with $\operatorname{char} k \neq 2$, X a normal variety of dimension n and $x \in X$ a point of multiplicity 2. Then its complete local ring can be written as*

$$\hat{\mathcal{O}}_{x,X} \cong k[[\mathbf{x}, y]]/(y^2 - r(\mathbf{x})) \cong k[[\mathbf{x}]][\sqrt{r(\mathbf{x})}].$$

Proof. By (2.48), $\hat{\mathcal{O}}_{x,X}$ is the normalization of a ring of the form

$$k[[\mathbf{x}, z]]/(z^2 + p_1(\mathbf{x})z + p_2(\mathbf{x})).$$

Since $\operatorname{char} k \neq 2$, we can complete the square and obtain the simpler form

$$k[[\mathbf{x}, z]]/(z^2 - p_3(\mathbf{x})).$$

Write $p_3\mathbf{x} = r(\mathbf{x})s(\mathbf{x})^2$, where $r(\mathbf{x})$ is square free. Then $y := zs^{-1}$ is integral, and we get the simpler ring

$$k[[\mathbf{x}, y]]/(y^2 - r(\mathbf{x})).$$

The singular set of $(y^2 - r(\mathbf{x}) = 0)$ is given by the equations

$$y = r = \partial r/\partial x_1 = \cdots = \partial r/\partial x_n = 0,$$

which is also the singular set of the $(n-1)$-dimensional hypersurface $(y = r = 0)$. Since r has no multiple factors, $\operatorname{Sing}(y = r = 0)$ has dimension $\leq n-2$, and so $\operatorname{Sing}(y^2 - r(\mathbf{x}) = 0)$ also has dimension $\leq n-2$.

Thus $k[[\mathbf{x}, y]]/(y^2 - r(\mathbf{x}))$ is normal by Serre's criterion ([Mat70, 17.I] or [Mat89, 23.8]), and hence, $\hat{\mathcal{O}}_{x,X} \cong k[[\mathbf{x}, y]]/(y^2 - r(\mathbf{x}))$. $\quad\square$

2.50 (Resolving $R[\sqrt{r}]$, $\operatorname{char} k \neq 2$). Let k be an algebraically closed field and $\operatorname{char} k \neq 2$. Let (R, m) be a regular local k-algebra of dimension 2 such that strong embedded resolution (1.73) holds for $\operatorname{Spec} R$. (For instance, R can be the local ring of a smooth point of a k-variety or the power series ring $k[[x, y]]$.)

Use (1.73) for the curve $C := (r = 0)$ in the smooth surface $S := \operatorname{Spec} R$ to get a projective birational morphism $f : S' \to S$ such that $C' := (f^*r = 0) \subset S'$ has simple normal crossings only. The pullback of the surface $S[\sqrt{r}]$ gives the surface $S'[\sqrt{f^*r}]$. Let $\bar{S} \to S'[\sqrt{f^*r}]$ denote its normalization.

In order to determine the singularities of $\bar{S}$, pick any closed point $s \in S'$. Since C' has simple normal crossings only, there are local coordinates $x, y \in m_{s'}$ such that $f^*r = x^a y^b(\text{unit})$. Thus over a neighborhood of $s \in S'$, we can write $\bar{S}$ as the normalization of

$$\mathcal{O}_{s,S'}[z']/(z'^2 - x^a y^b u),$$

where u is a unit. Set

$$z := z'x^{-\lfloor a/2 \rfloor}y^{-\lfloor b/2 \rfloor}, \quad c := a - 2\lfloor a/2 \rfloor, \quad d := b - 2\lfloor b/2 \rfloor.$$

Then $0 \leq c, d \leq 1$, z is integral over $\mathcal{O}_{s,S'}$ and $\bar{S}$ is also the normalization of

$$\mathcal{O}_{s,S'}[z]/(z^2 - x^c y^d u).$$

This ring is regular if $c = 0$ or $d = 0$. Indeed, if $c = 0, d = 0$, then we have one or two maximal ideals generated by (x, y); if $c = 0, d = 1$, then we have one maximal ideal generated by (x, z); and if $c = 1, d = 0$, then we have one maximal ideal generated by (z, y).

If $c = d = 1$, then $\mathcal{O}_{s,S'}[z]/(z^2 - xyu)$ is singular. However, if we blow up the origin once, we get a nonsingular surface. In the three charts we have

$$\mathcal{O}_{s,S'}[z_1]/(z_1^2 - y_1 u), \quad \mathcal{O}_{s,S'}[z_2]/(z_2^2 - x_2 u), \quad \mathcal{O}_{s,S'}[z_3]/(1 - x_3 y_3 u).$$

The first two of these cover the blow-up, and we can eliminate y_1 (resp., x_2) to see that the quotients are regular.

Thus $\bar{S}$ may be singular, but blowing up each singular point once results in a nonsingular surface. $\square$

REMARK 2.51. In characteristic 2 we still know that the completion of the local ring of a double point can be written as the normalization of a ring $R[z]/(z^2 + r_1 z + r_2)$.

In order to resolve such a singularity, a natural idea is to apply embedded resolution to the pair $C := (r_1 r_2 = 0) \subset \operatorname{Spec} R$. Just as in (2.50), we are then reduced to studying the normalization of rings of the form

$$R'[z]/(z^2 + x^a y^b(\text{unit}) + x^c y^d(\text{unit})).$$

If $c \geq 2a \geq 2$ or if $d \geq 2b \geq 2$, then we can replace z by $zx^{-a}y^{-b}$ to get simpler normal forms. In general, we can reduce to the special cases where $a = 0$ or $c \leq 1$ and $b = 0$ or $d \leq 1$.

This leaves quite a few cases open, but the main problem is that very innocent-looking polynomials produce complicated singularities. Let us just see some examples.

$z^2 + xz + (1 + x + y^3)$. The ramification locus is $x(1 + x + y^3) = 0$, which is smooth near the origin. Despite this, we get a singularity as seen by rewriting the equation as $(z + 1)^2 + x(z + 1) + y^3$, which is singular at $x = y = z + 1 = 0$.

$z^2 + x^3 z + y^4$. This is not even normal, but we need a coordinate change to see this. Indeed, we can rewrite it as $(z + y^2)^2 + x^3(z + y^2) + x^3 y^2$. Thus we can introduce $z_1 := (z + y^2)x^{-1}$ to get a new equation $z_1^2 + x^2 z_1 + xy^2$, which is now normal.

It is possible to get a good description of the situation, but it is quite complicated. See [Abh66] for details.

ASIDE 2.52 (Local rings of small multiplicity). The Albanese algorithm (2.33) shows that every surface is birational to a projective surface with points of multiplicity at most 2. This leads us to the study of low multiplicity points on algebraic varieties. We start by some examples showing what cannot be seen from the multiplicity alone.

(2.52.1) Let $S \subset \mathbb{A}^3$ be a surface, $s \in S$ a point and $C \subset \mathbb{A}^3$ any curve. Then by (2.71.3 and 4), $\text{mult}_s(S \cup C) = \text{mult}_s S$. More generally, embedded points and lower-dimensional irreducible components do not contribute to the multiplicity, and thus it is sensible to restrict to the pure dimensional case.

(2.52.2) For any $f \in (x,y)^2$, the surface $S = (z^2 = f(x,y)) \subset \mathbb{A}^3$ has multiplicity 2 at the origin. The coordinate ring of S is $R = k[x,y] + zk[x,y]$. Let us now consider the nonnormal ring

$$R_j := k[x,y] + z(x,y)^j \subset R.$$

As an exercise, show that R_j still has multiplicity 2 but its embedding dimension goes to infinity with j.

(2.52.3) Here we give an example of a Noetherian local integral domain of multiplicity 1, which is not regular.

Let K be a field and let $P = (x_1)$ and $Q = (x_2, x_3)$ be prime ideals in $K[x_1, x_2, x_3]$. Let R be the localization $R := K[x_1, x_2, x_3]_{P \cup Q}$. R has two maximal ideals PR and QR with residue fields $R/PR \cong K(x_2, x_3)$ and $R/QR \cong K(x_1)$.

Choose now $K = k(t_1, t_2, \dots)$, where the t_i are infinitely many algebraically independent variables over k. The point of this strange choice is that now there is a k-isomorphism $\phi : R/PR \cong R/QR$. We can use ϕ to "pinch together" the two closed points. That is, set

$$S := \{ f \in R : \phi(f \mod PR) = (f \mod QR) \} \subset R.$$

We see that R is the normalization of S, and thus R is not normal and, hence, certainly not regular. S and R both have dimension 2, and so $\text{mult}\, S = \text{mult}\, R_Q = 1$.

This could not have been done if we viewed R as a K-algebra, and S cannot be made into a K-algebra because there is no K-isomorphism between $K(x_2, x_3)$ and $K(x_1)$. The change from K-algebras to k-algebras means that S is not "geometric," meaning for instance that it is not the localization of a finitely generated k-algebra.

(2.52.4) Even in characteristic 2, the complete local ring of a normal double point is of the form

$$k[[x_1, \dots, x_n, y]] / (y^2 + r_1(\mathbf{x})y + r_2(\mathbf{x})),$$

but I do not know any elementary proof.

(2.52.5) An algebraic version of the Bertini bounds (2.40) says that the embedding dimension of a Cohen-Macaulay local ring R is bounded from above by $\dim R + \text{mult}\, R - 1$ (cf. [Sal79]).

I do not know what happens for normal rings. In particular, I do not know whether the embedding dimension of a normal triple point $x \in X$ is at most $2 + \dim X$.

2.7. Embedded resolution using Weierstrass' theorem

This section shows how the embedded resolution of curves using the Weierstrass theorem (1.89) can be generalized to surfaces.

Let $S \subset X$ be a reduced surface in a smooth 3-fold X. Our aim is to perform a sequence of blow-ups centered at points and smooth curves to obtain a smooth surface $S' \subset X'$ sitting inside a smooth 3-fold X'. We start with the highest multiplicity points and reduce the maximal multiplicity step-by-step.

DEFINITION 2.53. Let $S \subset X$ be a reduced surface in a smooth 3-fold X. The set of points $x \in X$, where $\text{mult}_x S \geq m$, is a closed subvariety, denoted by $\text{Sing}_m S$. (We consider it with reduced scheme structure.)

Let m be the maximum multiplicity of S. A *blow-up sequence* of order m starting with (X, S) is a sequence

$$\Pi : (X_r, S_r) \xrightarrow{\pi_{r-1}} (X_{r-1}, S_{r-1}) \xrightarrow{\pi_{r-2}} \cdots$$
$$\xrightarrow{\pi_1} (X_1, S_1) \xrightarrow{\pi_0} (X_0, S_0) = (X, S),$$

with the following properties.

(1) Each $\pi_i : X_{i+1} \to X_i$ is a blow-up with center Z_i, which is either a point in $\text{Sing}_m S_i$ or a smooth curve in $\text{Sing}_m S_i$.

(2) The S_i are defined recursively by the formula $S_{i+1} = B_{Z_i} S_i$. Alternatively, S_i is the birational transform of $S \subset X$ in X_i.

The main result of this section is the following.

THEOREM 2.54 (Multiplicity reduction). *Let $S \subset X$ be a reduced surface in a smooth 3-fold X defined over an algebraically closed field of characteristic zero. Let $m \geq 2$ be the maximum multiplicity of S. Then there is a blow-up sequence $\Pi : (X_r, S_r) \to (X, S)$ of order m such that the maximum multiplicity of S_r is less than m.*

By repeatedly applying (2.54) we eventually drop the multiplicity to 1 and obtain the following.

COROLLARY 2.55 (Embedded resolution). *Let $S \subset X$ be a reduced surface in a smooth 3-fold X defined over an algebraically closed field of characteristic zero. Then there is a composite of smooth blow-ups $\Pi : X' \to X$ such that the birational transform $S' \subset X'$ of S is smooth.* □

The key step is to understand that multiplicity reduction for $S \subset X$ follows from a variant of embedded resolution for curves. This is achieved, at least locally, using coefficient curves.

DEFINITION 2.56 (Maximal contact and coefficient curves). Let $S \subset X$ be a reduced surface in a smooth 3-fold X. Pick a point $0 \in S$, where

$\operatorname{mult}_0 S = m$. Choose suitable local coordinates x, y, z and apply the Weierstrass preparation theorem (1.93) to get an equation

$$z^m + a_1(x,y)z^{m-1} + \cdots + a_m(x,y) = 0.$$

We can kill the z^{m-1} term by a substitution $z \mapsto z - \frac{1}{m}a_1(x,y)$ to get another local equation

$$f := z^m + b_2(x,y)z^{m-2} + \cdots + b_m(x,y) = 0. \qquad (2.56.1)$$

Here $\operatorname{mult}_0 b_i \geq i$ since $\operatorname{mult}_0 S = m$. Note that we have to use either complex analytic or formal power series coordinates. The complex analytic case is easier to imagine since we still have many closed points left, but formal power series coordinates work just as well.

From now on we forcibly separate the z coordinate from the (x,y)-coordinates. That is, we fix z and also the coordinate projection to the plane $H := (z = 0)$. The equation in (2.56.1) will be the *normal form* that we aim to keep through the resolution process.

The smooth surface $(z = 0)$ will be a surface of *maximal contact*, and the curves (or rather curve germs) $B_i := (b_i = 0)$ are called the *coefficient curves* of S in the coordinate system (x, y, z).

Unfortunately, the coefficient curves depend on the coordinate system. (For instance, we can view the surface $u^3 + vw^2 + v^3$ as $z^3 + xy^2 + x^3$ or as $z^3 + zy^2 + x^3$.) For surfaces, we have to deal with only finitely many "special" points, and one can choose the local coordinate systems independently. Starting with dimension 3, however, one has to deal with serious compatibility problems among the various local coordinate systems. Solving this difficulty is one of the tasks accomplished in Chapter 3.

REMARK 2.57 (Maximal contact in positive characteristic). There are two places where we use the characteristic zero assumption.

First, in order to get the normal form (2.56.1), we use a substitution $z \mapsto z - \frac{1}{m}a_1$. This is only possible when the characteristic does not divide m.

Second, in proving (2.59.1) we use that certain numerical coefficients coming from differentiating $(m-1)$-times are not zero. This holds if m is less than the characteristic.

Thus (2.54) also holds whenever the characteristic is larger than the multiplicity.

The key idea of maximal contact can be summarized as follows.

PRINCIPLE 2.58. *Multiplicity reduction for $S \subset X$ near $s \in S$ is equivalent to a modified version of embedded resolution for $B_2, \ldots, B_m \subset H := (z = 0)$.*

2.59 (Local computations). Before any blow-ups, note that one can read off $\operatorname{Sing}_m S$ from the B_i.

Claim 2.59.1. $f = z^m + b_2(x, y)z^{m-2} + \cdots + b_m(x, y)$ has multiplicity $\geq m$ at a point p iff $z(p) = 0$ and $\text{mult}_p\, b_i \geq i$ for every i.

Proof. A polynomial has multiplicity $\geq m$ at a point p iff all the $(m-1)$st partials vanish at p. In particular

$$\frac{\partial^{m-1} f}{\partial z^{m-1}}(p) = m! \cdot z(p) = 0.$$

Thus all points of multiplicity $\geq m$ of S are on the plane $(z = 0)$.

After restricting to $(z = 0)$, the other $(m-1)$st partials of f can be written as

$$\frac{\partial^{m-1}(F, m)}{\partial x^i \partial y^j \partial z^{m-1-i-j}}\big|_{(z=0)} = (m - i - j)! \cdot \frac{\partial^{i+j} b_{i+j+1}(x, y)}{\partial x^i \partial y^j}.$$

All of these vanish at p iff $\text{mult}_p\, b_{i+j+1} \geq i + j + 1$ for every i, j. □

Next we investigate how the normal form changes when we blow up points and curves.

2.59.2 *Point blow-up.* Let us blow up the point 0 to get $\pi : B_0 X \to X$, and consider the chart $x_1 = x, y_1 = y/x, z_1 = z/x$. The equation of $B_0 S$ is

$$F := z_1^m + x_1^{-2} b_2(x_1, y_1 x_1)z_1^{m-2} + \cdots + x_1^{-m} b_m(x_1, y_1 x_1). \qquad (2.59.2.x)$$

We get a similar equation in the chart $x_2 = x/y, y_2 = y, z_2 = z/y$:

$$z_2^m + y_2^{-2} b_2(x_2 y_2, y_2)z_2^{m-2} + \cdots + y_2^{-m} b_m(x_2 y_2, y_2) = 0, \qquad (2.59.2.y)$$

while in the third chart $x_3 = x/z, y_3 = y/z, z_3 = z$, we get

$$1 + z_3^{-2} b_2(x_3 z_3, y_3 z_3)z_3^{m-2} + \cdots + z_3^{-m} b_m(x_3 z_3, y_3 z_3) = 0. \qquad (2.59.2.z)$$

The last equation $(2.59.2.z)$ is of the form $1 + g(x_3, y_3, z_3) = 0$, where $g \in (x_3, y_3)$; hence, the origin $(0, 0, 0)$ is not a solution. This means that the first two charts $(2.59.2.x\text{-}y)$ cover the whole birational transform of S, and we use only these in the sequel. These in turn are completely symmetrical, and hence, it is enough to see what happens in the first chart $(2.59.2.x)$.

We can apply $(2.59.1)$ to the equation $(2.59.2.x)$ to conclude that all points of multiplicity $\geq m$ of $B_0 S$ are on the birational transform of the hyperplane $(z = 0)$. Hence all the points on the exceptional divisor of π, where $B_0 S$ has multiplicity $\geq m$, are on the curve $(z_1 = x_1 = 0)$. By a coordinate change $y_1 \mapsto y_1 + c$, any such point can be made into the origin and the equation $(2.59.2.x)$ stays in our normal form. This coordinate change corresponds to $y \mapsto y + cx$, and hence it keeps the original separation of the z and (x, y)-coordinates.

2.59.3 *Curve blow-up.* Let $0 \in C$ be a smooth curve (germ) along which S has multiplicity $\geq m$. Then $C \subset H := (z = 0)$, and by changing the x, y coordinates we can assume that $C = (z = x = 0)$. This means that x^i divides b_i for every i.

Let us blow up C to get $\pi : B_C X \to X$. In the chart $x_1 = x, y_1 = y, z_1 = z/x$, the equation of $B_C S$ is

$$F := z_1^m + x_1^{-2} b_2(x_1, y_1) z_1^{m-2} + \cdots + x_1^{-m} b_m(x_1, y_1), \qquad (2.59.3.x)$$

and in the chart $x_3 = x/z, y_3 = y, z_3 = z$, we get

$$1 + z_3^{-2} b_2(x_3 z_3, y_3 z_3) z_1^{m-2} + \cdots + z_3^{-m} b_m(x_3 z_3, y_3 z_3) = 0. \qquad (2.59.3.z)$$

As before we see that the first chart covers $B_C S$.

For both types of blow-ups, the normal form (2.56.1) is preserved, and every further blow-up is computed the same way. By induction, we obtain the following.

Claim 2.59.4. Let X be a smooth 3-fold and $S \subset X$ a reduced surface. Pick a point $s \in S$ and a local (analytic or formal) neighborhood $x \in X_0$ such that $S_0 := X_0 \cap S$ is defined by a local equation

$$z^m + b_2(x, y) z^{m-2} + \cdots + b_m(x, y) = 0.$$

Further, let

$$\Pi : (X_r, S_r) \xrightarrow{\pi_{r-1}} (X_{r-1}, S_{r-1}) \xrightarrow{\pi_{r-2}} \cdots$$
$$\xrightarrow{\pi_1} (X_1, S_1) \xrightarrow{\pi_0} (X_0, S_0) \subset (X, S)$$

be a smooth blow-up sequence of order m starting with (X, S). Then we have the following.

(i) All points of all the S_j have multiplicity $\leq m$.
(ii) All points of multiplicity $= m$ of S_j are on the birational transform $H_j \subset X_j$ of the plane $H_0 := (z = 0)$.
(iii) For each point $s \in \text{mult}_m S_j$ we have specified local coordinates (x_s, y_s, z_s) such that the equation of S_j is in normal form.
(iv) There are globally defined coefficient curves $B_{i,j} \subset H_j$ that agree with the locally defined coefficient curves in every coordinate system.

Moreover, the coefficient curves are computed by the following rules.

(v) (Point blow-up) Let $\pi_j : H_{j+1} \to H_j$ be the blow-up of $s_j \in H_j$ and $E_{j+1} \subset H_{j+1}$ the exceptional curve. Let $(\pi_j)_*^{-1} B_{i,j} \subset H_{j+1}$ denote the birational transform of $B_{i,j}$. Then

$$B_{i,j+1} = (\pi_j)_*^{-1} B_{i,j} + (\text{mult}_{s_j} B_{i,j} - i) E_{j+1}.$$

(vi) (Curve blow-up) Let $\pi_j : H_{j+1} \to H_j$ be the blow-up of $C_j \subset H_j$. Then $H_{j+1} = H_j$ and

$$B_{i,j+1} = B_{i,j} - i C_j. \qquad \square$$

REMARK 2.60. The maximal contact surfaces H_j exhibit a curious lo-cal/global nature. We start with the original choice H_0, which is only locally defined near the point $s \in S$. After blow-ups, however, we do not make new local choices but always use the birational transform of the orig-inal H_0. So H behaves like a local object on X but as a global object with respect to blow-up sequences.

For surfaces this double nature can be avoided, but in higher dimensions this seems to be necessary. See (3.6) for one aspect of this phenomenon.

ALGORITHM 2.61 (Reduction to monomial coefficients).

Step 0 (Preparing local charts). Let $S \subset X$ be a reduced surface inside a smooth 3-fold, all defined over an algebraically closed field of characteristic zero. Let m be the maximum multiplicity of S. Let $s^1, \dots, s^t \in S$ be points such that either $\mathrm{Sing}_m S$ is 1-dimensional and singular at s^i, or s^i is an isolated point of $\mathrm{Sing}_m S$. Throw in a few other a points to achieve that every irreducible component of $\mathrm{Sing}_m S$ contains at least one point s^j. These are "special" points of S, and we have to pay close attention to these in the resolution.

For each such s^j choose local (analytic or formal) coordinates x_j, y_j, z_j such that near s^j the surface S can be given by an equation

$$z_j^m + b_2^{(j)}(x_j, y_j) z_j^{m-2} + \cdots + b_m^{(j)}(x_j, y_j) = 0 \qquad (2.61.1)$$

with coefficient curves

$$B_i^j := (z_j = b_i^{(j)} = 0) \subset H^j := (z_j = 0).$$

Note that S is an actual surface, but H^j is only a germ of a surface around s^j and the coefficient curves B_i^j are germs of curves.

Remark 2.61.2. These are not the only points that one needs to look at carefully during resolution, and it would have been advantageous to specify the right set of "special" points in advance.

The following example shows that even the ideal of $(m-1)$st derivatives does not always tell which points of $\mathrm{Sing}_m S$ need special attention. Let $S := (z^3 + zx^8 + x^9 y^3 = 0)$. The ideal of the triple locus is (z, x^7), and the origin is not a "special" point. After blowing up $(z = x = 0)$ three-times, we get the equation

$$(z/x^3)^3 + (z/x^3)x^2 + y^3 = 0,$$

which has an isolated triple point at the origin.

It seems easier to repeat a loop of the algorithm twice rather than to worry about the correct selection of "special" points at the beginning.

Step 1 (Making the coefficient curves normal crossing). By performing point blow-ups over each $s^1, \dots, s^t$, we get $\Pi : X_r \to X$ and $\Pi : S_r \to S$

such that over each $s^j \in S$ the coefficient curves $B_{i,r}^j$ cross normally at every point of $\mathrm{Sing}_m S_r \cap \Pi^{-1}(s^j)$.

Indeed, by (2.59.1), for $p \in \Pi^{-1}(s^j)$ we have $\mathrm{mult}_p B_{i,r}^j \geq i$ for every i iff $\mathrm{mult}_p S_r = m$. In particular, every blow-up we make over s^j in the blow-up sequence $\Pi : X_r \to X$ is also a possible blow-up in the embedded resolution for $(B_2^j + \cdots + B_m^j \subset H^j)$. Thus by (1.73) we have to stop after finitely many such blow-ups.

In particular the 1-dimensional part of $\mathrm{Sing}_m S_r$ is a curve with only ordinary nodes. Since we have worked with local analytic equations, we cannot tell if a node is an intersection of two different irreducible components or the self-intersection of one irreducible component. However, if we blow up each node one more time, we get $S_{r+1} \to S_r$ and $\mathrm{Sing}_m S_{r+1}$ is a union of smooth curves, which intersect transversally.

Step 2 (Blowing up m-fold curves). We blow up first the Π-exceptional m-fold curves. With each such blow-up, the multiplicity of the coefficient curve B_i^j decreases by i, and so eventually we run out of Π-exceptional m-fold curves.

Next we look at the birational transform of $\mathrm{Sing}_m S$, which is smooth by Step 1.

Let $C \subset \mathrm{Sing}_m S_{r+1}$ be an irreducible component. Then $\Pi(C) \subset S$ may appear on several of the local charts H^j, and it is not obvious that on different charts it appears with the same multiplicity in the local coefficient curves. This, however, does not matter. As we blow up C, the multiplicity of any local coefficient curve B_i^j contained in C decreases by i at each step. So the procedure eventually stops. Moreover, the blow-ups stop when the multiplicity of the birational transform of S_{r+1} along the birational transform of C drops below m, so it does not matter which chart we consider.

After all these blow-ups we get $\Pi' : X_{r'} \to X$ such that $\mathrm{Sing}_m S_{r'}$ is zero-dimensional. There are two types of points in $\mathrm{Sing}_m S_{r'}$.

First, as we blow up the curves corresponding to $\mathrm{Sing}_m S$, we eventually eliminate all m-fold curves but may have some m-fold points left over that do not lie on any $H_{r'}^j$, as in (2.61.2). We have made no preparation to deal with these "accidental" points. For these points we have to repeat all the steps. Thus we can reduce these to the next case. Second, there may be m-fold points on the maximal contact surfaces $H_{r'}^j$. By construction, the coefficient curves form a simple normal crossing divisor.

We have reached the end of the first reduction step, and it is time to codify the resulting class of singularities.

DEFINITION 2.62 (Monomial coefficients). We say that $f(x, y, z)$ has *monomial coefficients* if

$$f(x, y, z) = z^m + b_2(x, y)z^{m-2} + \cdots + b_m(x, y),$$

where $b_i = x^{a_i} y^{c_i} u_i$ and the u_i are units. Geometrically this is equivalent to assuming that the local coefficient curves $B_2 + \cdots + B_m$ form a normal crossing divisor.

Note that the origin is an isolated m-fold point of $(f = 0)$ iff

    (1) $a_i + c_i \geq i$ for every i,
    (2) $a_j < j$ for some j, and
    (3) $c_j < j$ for some j.

REMARK 2.63. The Jungian method discussed in Section 2.3 reduced the resolution of arbitrary surface singularities to the resolution of singularities of the form $(z^m = x^a y^b)$. More intrinsically put, these are the singularities that admit a projection to a plane whose branch locus is a normal crossing divisor (cf. [Lip83]).

In (2.62) we allow other monomial coefficients as well; this leads to much more complicated singularities. The branch divisor of the coordinate projection to the (x, y)-plane is given by the discriminant of f. For $m \geq 3$, the discriminant of a polynomial with monomial coefficients need not be a normal crossing divisor. For instance, the discriminant of $z^3 + xz + y$ is the cuspidal curve $27y^2 + 4x^3 = 0$.

2.64 (Blow-up formulas). Consider

$$z^m + x^{a_2} y^{c_2} u_2 z^{m-2} + \cdots + x^{a_m} y^{c_m} u_m.$$

What happens under a multiplicity m blow-up?

If we blow up the origin, then in the chart $(x, \frac{y}{x}, \frac{z}{x})$ we get

$$\left(\frac{z}{x}\right)^m + x^{a_2+c_2-2}\left(\frac{y}{x}\right)^{c_2} u_2 \left(\frac{z}{x}\right)^{m-2} + \cdots + x^{a_m+c_m-m}\left(\frac{y}{x}\right)^{c_m} = 0,$$

and symmetrically in the $(\frac{x}{y}, y, \frac{z}{y})$ chart. These two charts cover the whole blow-up.

If we blow up $(x = z = 0)$, then in the chart $(x, y, \frac{z}{x})$ we get

$$\left(\frac{z}{x}\right)^m + x^{a_2-2} y^{c_2} u_2 \left(\frac{z}{x}\right)^{m-2} + \cdots + x^{a_m-m} y^{c_m} = 0$$

and this chart covers the whole blow-up.

Before we go on to resolve these, we should look at some examples.

EXAMPLE 2.65. It can easily happen that the blow-up has an m-fold singularity along the exceptional curve. For instance, for

$$z^4 + xy^9 z^2 + x^9 yz + x^5 y^5 \qquad\qquad (2.65.1)$$

the z^2-term shows that its multiplicity along $(x = z = 0)$ is < 4 and the z-term shows that its multiplicity along $(y = z = 0)$ is < 4. After one blow-up, in the $(x_1 = x, y_1 = x/y, z_1 = z/x)$-chart we get

$$z_1^4 + x_1^8 y_1^9 z_1^2 + x_1^7 y_1 z_1 + x_1^6 y_1^5, \qquad (2.65.2)$$

which has multiplicity 4 along $x_1 = z_1 = 0$. If we blow up the exceptional curve, we get in the $x_2 = x_1, y_2 = y_1, z_2 = z_1/x_1$ coordinates

$$z_2^4 + x_2^6 y_2^9 z_2^2 + x_2^4 y_2 z_2 + x_2^2 y_2^5, \qquad (2.65.3)$$

which now has an isolated m-fold point, but along $(x = z = 0)$ this is shown by the constant term. This makes it hard to keep track of improvements. Indeed, is (2.65.3) any better than (2.65.1)?

Everything is much easier if one of the terms is smallest in both x and y powers, but we have to be careful with this. Start with

$$z^4 + x^3 y^3 z^2 + x^6 y^3. \qquad (2.65.4)$$

Here the z^2-term is the smallest in both x and y powers. After one blow-up we get

$$z_1^4 + x_1^4 y_1^3 z_1^2 + x_1^5 y_1^3, \qquad (2.65.5)$$

followed by

$$z_2^4 + x_2^5 y_2^3 z_2^2 + x_2^4 y_2^3, \qquad (2.65.6)$$

where now the constant term has the smallest x and y powers.

The problem is that in the blow-up we transform the z^j-terms by the rule

$$x^{a_j} y^{c_j} \mapsto x^{a_j + c_j - j} y^{c_j},$$

which depends on j. The following definition corrects this problem.

DEFINITION 2.66. We say that the z^j term dominates

$$f(x, y, z) = z^m + b_2 z^{m-2} + \cdots + b_m, \quad \text{where } b_i = x^{a_i} y^{c_i} u_i$$

if $a_j / j \le a_i / i \ \forall i$ and $c_j / j \le c_i / i \ \forall i$.

From the blow-up formulas (2.64) we read off the following.

LEMMA 2.67. *Assume that* $z^m + x^{a_2} y^{c_2} u_2 z^{m-2} + \cdots + x^{a_m} y^{c_m} u_m$ *is dominated by its* z^j *term. After any sequence of blow-ups of points or curves of multiplicity* m, *the resulting local equations are all dominated by their* z^j *term.* □

ALGORITHM 2.68 (Multiplicity reduction for monomial coefficients). We start with $S \subset X$, where S has only finitely many points s^ℓ of maximal multiplicity m and at each s^ℓ it has a local equation with monomial coefficients

$$z_\ell^m + b_2^{(\ell)}(x_\ell, y_\ell) z^{m-2} + \cdots + b_m^{(\ell)}(x_\ell, y_\ell), \quad \text{where } b_i^{(\ell)} = x_\ell^{a_{i\ell}} y_\ell^{c_{i\ell}} u_{i\ell}.$$

We deal with the points separately, and so we drop the index ℓ from the notation.

Step 1 (Reduction to dominated equations). From the blow-up formulas (2.64) we see that under a blow-up the sequence of exponents undergoes a transformation

$$(a_i, c_i) \mapsto (a_i + c_i - i, c_i) \quad \text{and} \quad (a_i, c_i) \mapsto (a_i, a_i + c_i - i).$$

These look similar to the steps in (1.79). A key difference is that here the pairs (a_i, c_i) are transformed by a rule that depends on i. In order to get rid of this dependence, we can switch to the pairs

$$\left(\tfrac{a_2}{2}, \tfrac{c_2}{2}\right), \dots, \left(\tfrac{a_m}{m}, \tfrac{c_m}{m}\right)$$

to get the transformation rules

$$\left(\tfrac{a_i}{i}, \tfrac{c_i}{i}\right) \mapsto \left(\tfrac{a_i}{i} + \tfrac{c_i}{i} - 1, \tfrac{c_i}{i}\right) \quad \text{and} \quad \left(\tfrac{a_i}{i}, \tfrac{c_i}{i}\right) \mapsto \left(\tfrac{a_i}{i}, \tfrac{a_i}{i} + \tfrac{c_i}{i} - 1\right).$$

Now we can apply (1.79) with $k_s = -1$, and after finitely many blow-ups we get $\Pi : X_r \to X$ such that at every point of multiplicity m of S_r the local equation

$$z^m + x^{a_2} y^{c_2} u_2 z^{m-2} + \cdots + x^{a_m} y^{c_m} u_m$$

is dominated by its z^j term for some j (where the a_i, c_i and j all depend on the point).

Step 2 (Blowing up m-fold curves). Exactly as in Step 2 of (2.61), we blow up the Π-exceptional m-fold curves. With each such blow-up, the multiplicity of the coefficient curve $B_{i,t}$ decreases by i, and so eventually we run out of Π-exceptional m-fold curves.

By (2.67) now we have only finitely many m-fold points, and at each of them one of the z^j terms dominates.

Step 3 (Blowing up m-fold points). Assume that we have an isolated m-fold point

$$z^m + x^{a_2} y^{c_2} u_2 z^{m-2} + \cdots + x^{a_m} y^{c_m} u_m$$

and the z^j-term dominates. Thus $a_j < j$, $b_j < j$ and $2j > a_j + b_j \geq j$. After a blow-up we get

$$z^m + x^{a_2 + c_2 - 2} y^{c_2} u_2 z^{m-2} + \cdots + x^{a_m + c_m - m} y^{c_m} u_m.$$

The dominant term went from $x^{a_j} y^{c_j} u_j$ to $x^{a_j + c_j - j} y^{c_j} u_j$. Observe that

$$a_j + c_j - j + c_j = a_j + c_j - (j - c_j) < a_j + c_j.$$

Thus after at most j blow-ups the multiplicity of the dominant z^j-term drops below j, and so the multiplicity of the surface drops below m. We are done. $\qquad\square$

2.8. Review of multiplicities

If $X = f(x_1, \ldots, x_n) \subset \mathbb{A}^n$ is a hypersurface, then its *multiplicity* at the origin is the lowest degree of monomials occurring in f. How do we define the multiplicity of an arbitrary algebraic variety or a scheme at a point? The answer is not at all clear, but there are various possibilities. Here are three of them.

Let $X \subset \mathbb{A}^n$ be an affine scheme of dimension d and $x \in X$ a point. The following are three possible ways to define the multiplicity $\mathrm{mult}_x X$.

(1) Take a general curve section $X \cap H_1 \cap \cdots \cap H_{d-1}$, and consider its multiplicity.
(2) Take a general projection $\pi : X \to \mathbb{A}^{d+1}$, and consider the multiplicity of $\pi(X)$ at $\pi(x)$.
(3) Take the tangent cone of X at x, and consider its degree.

It turns out that these three versions are equivalent, and this is not very difficult to prove. So we are clearly on the right track.

Note, however, that all three approaches present difficulties when we want to compute the multiplicity.

In the first two cases one really needs general hyperplanes or a general projection. For instance, $x^2 + y^3 + z^4 = 0$ has multiplicity 2 at the origin, but its intersection with the $(x = 0)$-plane

$$(x^2 + y^3 + z^4 = 0) \cap (x = 0) \cong (y^3 + z^4 = 0) \subset \mathbb{A}^2$$

has multiplicity 3 at the origin. Similarly, the curve given parametrically by $t \mapsto (t^2, t^3, t^4)$ has multiplicity 2 at the origin, but the projection to the last two coordinates gives $t \mapsto (t^3, t^4)$ and a multiplicity 3 curve. General equations are hard to write down, and this makes the first two variants sometimes hard to use in practice.

The degree of the tangent cone is easy to compute once we know what the tangent cone is. It is the variety defined by the leading terms (in their Taylor expansion around x) of all elements of the ideal of X. The trouble is that usually the ideal of X is given by generators $I_X = (f_1, \ldots, f_t)$, and it is not enough to use the leading terms of the f_i's. If

$$\sum (\text{leading term of } g_i) \cdot (\text{leading term of } f_i) = 0,$$

then the leading term of $\sum g_i f_i$ depends on the higher-order terms of the g_i and the f_j.

It turns out that we get a more flexible theory if we define the multiplicity for any sheaf or module and not just at a point but with respect to any ideal sheaf that defines a 0-dimensional subscheme.

DEFINITION 2.69 (Hilbert-Samuel multiplicity). Let R be a ring and $I \subset R$ an ideal such that R/I is Artinian. Equivalently, I defines a 0-dimensional subscheme in Spec R. Let M be a finite R-module. Then

$$\mathrm{gr}_I M := \sum_{s \geq 0} \mathrm{gr}_I^s M := \sum_{s \geq 0} I^s M / I^{s+1} M$$

is a finite $\mathrm{gr}_I R$-module, and so by [AM69, 11.2] there is a unique polynomial $H(\mathrm{gr}_I M, s)$ such that

$$\mathrm{length}_R \mathrm{gr}_I^s M = H(\mathrm{gr}_I M, s) \quad \text{for } s \gg 1.$$

This implies that there is a polynomial $HS_I(M, s)$ such that

$$\mathrm{length}_{R/m} M / I^{s+1} M = HS_I(M, s) \quad \text{for } s \gg 1.$$

It is clear that

$$H(\mathrm{gr}_I M, s) = HS_I(M, s) - HS_I(M, s - 1).$$

The operation of going from $HS_I(M, s)$ to $H(\mathrm{gr}_I M, s)$ could be thought of as a discrete version of differentiation. Thus the leading coefficients of $HS_I(M, s)$ and of $H(\mathrm{gr}_I M, s)$ determine each other, except when R is Artinian, that is, when $\deg HS_I(M, s) = 0$.

The polynomial $HS_I(M, s)$ is called the *Hilbert-Samuel polynomial* of M with respect to I. As in [AM69, 11.2] one easily checks that

$$\deg H(\mathrm{gr}_I M, s) \leq \dim R - 1 \quad \text{and} \quad \deg HS_I(M, s) \leq \dim R.$$

We define the *multiplicity* of M with respect to I to be

$$\mathrm{mult}_I(M) := (\dim R)! \cdot \big(\text{the coefficient of } s^{\dim R} \text{ in } HS_I(M, s)\big).$$

If $\dim R \geq 1$ then we can also compute it as

$$\mathrm{mult}_I(M) = (\dim R - 1)! \cdot \big(\text{the coefficient of } s^{\dim R - 1} \text{ in } H(\mathrm{gr}_I M, s)\big).$$

(In the literature, the multiplicity is also frequently denoted by $e(I, M)$ or by $\mu(I, M)$.) It is easy to prove directly from the definition that the multiplicity is a natural number.

EXAMPLE 2.70. (1) Let $R = k[x_1, \ldots, x_n]$ and $m = (x_1, \ldots, x_n)$, or more generally, (R, m) can be a regular local ring with maximal ideal m and $\dim_{R/m} m/m^2 = n$. Then

$$\mathrm{gr}_m^s R \cong (\text{degree } s \text{ homogeneous polynomials}),$$

and thus

$$H(\mathrm{gr}_m R, s) = \binom{s + n - 1}{n - 1} \quad \text{and so} \quad \mathrm{mult}_m(R) = 1.$$

(2) For a hypersurface point, we get back the usual notion of multiplicity. Indeed, here $S = R/f$, where (R, m) is a regular local ring with

maximal ideal $m = (x_1, \ldots, x_n)$ and $f \in m^d \setminus m^{d+1}$ with leading term f_d. Then

$$\operatorname{gr}_m^s S \cong \operatorname{gr}_m^s R / f_d \operatorname{gr}_m^{s-d} R,$$

and thus

$$H(\operatorname{gr}_m S, s) = \binom{s+n-1}{n-1} - \binom{s-d+n-1}{n-1} \quad \text{and so} \quad \operatorname{mult}_m(S) = d.$$

(3) If R is 0-dimensional, that is, Artinian, then M has finite length and $I^s M = 0$ for $s \gg 1$. Thus $\operatorname{mult}_I M = \operatorname{length} M$ is independent of I.

The basic properties of the multiplicity are gathered in the following lemma.

LEMMA 2.71. *Let R be a ring, $I \subset R$ an ideal such that R/I is Artinian and M, M_i finite R-modules.*

   (1) *If $I \supset J$, then $\operatorname{mult}_I M \leq \operatorname{mult}_J M$.*
   (2) *If $I = \cap I_j$ and $I_j + I_k = R$ for $j \neq k$, then $\operatorname{mult}_I M = \sum \operatorname{mult}_{I_j} M$.*
   (3) *If $\dim \operatorname{Supp} M < \dim R$, then $\operatorname{mult}_I(M) = 0$.*
   (4) *If $0 \to M_1 \to M_2 \to M_3 \to 0$ is exact, then $\operatorname{mult}_I(M_2) = \operatorname{mult}_I(M_1) + \operatorname{mult}_I(M_3)$.*
   (5) *If R is an integral domain, then $\operatorname{mult}_I(M) = \operatorname{mult}_I(R) \cdot \operatorname{rank}_R M$.*
   (6) *If S/R is finite, $\dim S = \dim R$ and M is a finite S-module, then we have $\operatorname{mult}_{IS}(M) = \operatorname{mult}_I(M)$.*

Proof. (1) is clear. If $I = \cap I_j$ and $I_j + I_k = R$ for $j \neq k$, then by the Chinese remainder theorem $M/I^{s+1}M \cong \sum_j M/I_j^{s+1}M$, proving (2).

If $\dim \operatorname{Supp} M < \dim R$, then M is also a module over the smaller dimensional $R/\operatorname{Ann}(M)$, and hence $\deg HS_I(M, s) < \dim R$, giving (3).

The exact sequence $0 \to M_1 \to M_2 \to M_3 \to 0$ gives

$$0 \to M_1/(M_1 \cap I^s M_2) \to M_2/I^s M_2 \to M_3/I^s M_3 \to 0.$$

Clearly $I^s M_1 \subset M_1 \cap I^s M_2$, and by the Artin-Rees lemma (cf. [AM69, 10.9]), there is an s_0 such that $M_1 \cap I^s M_2 \subset I^{s-s_0} M_1$ for every $s \geq s_0$. This gives (4) right away.

We prove (5) by induction on the rank. If $\operatorname{rank}_R M = 0$ then M is a torsion module, and so $\operatorname{mult}_I(M) = 0$ by (3). If $\operatorname{rank}_R M \geq 1$, then take any injection $j : R \hookrightarrow M$. Using (4) and induction we see that

$$\begin{aligned}
\operatorname{mult}_I M &= \operatorname{mult}_I(R) + \operatorname{mult}_I\big(M/j(R)\big) \\
&= \operatorname{mult}_I(R) + (\operatorname{rank}_R M - 1)\operatorname{mult}_I(R) = \operatorname{rank}_R M \cdot \operatorname{mult}_I(R).
\end{aligned}$$

Let S/R be a finite ring extension and M a finite S-module. Then $I^s M = (IS)^s M$, and so the Hilbert-Samuel polynomial of M as an R-module is the same as the Hilbert-Samuel polynomial of M as an S-module. This implies (6). $\square$

We can translate these results to the language of varieties to see how multiplicities behave under finite morphisms.

THEOREM 2.72. *Let Y be an irreducible and reduced variety and X a scheme, both of pure dimension d. Let $f : X \to Y$ be a finite morphism, $y \in Y$ a closed point and $I \subset \mathcal{O}_{y,Y}$ an m_y-primary ideal. Then we have the following.*

(1) $\sum_{x \in f^{-1}(y)} \operatorname{mult}_{I\mathcal{O}_{x,X}}(\mathcal{O}_{x,X}) = \deg f \cdot \operatorname{mult}_I(\mathcal{O}_{y,Y})$.

(2) $\sum_{x \in f^{-1}(y)} \operatorname{mult}_x X \leq \deg f \cdot \operatorname{mult}_y Y$.

(3) *Let $f : Z \to Y$ be the normalization of Y. Then $\operatorname{mult}_z Z \leq \operatorname{mult}_{f(z)} Y$ for every closed point $z \in Z$.*

Proof. The question is local, and so we may assume that $Y = \operatorname{Spec} R$ is affine. Then $X = \operatorname{Spec} S$ is also affine. By (2.71.2 and 6)

$$\deg f \cdot \operatorname{mult}_I(\mathcal{O}_{y,Y}) = \operatorname{mult}_{IS}(S) = \sum_{x \in f^{-1}(y)} \operatorname{mult}_{I\mathcal{O}_{x,X}}(\mathcal{O}_{x,X}).$$

Applying (1) to $I = m_y$ and using that $\operatorname{mult}_{m_x}(S_x) \leq \operatorname{mult}_{IS_x}(S_x)$ gives (2) which implies (3) directly. $\qquad\square$

In order to compute multiplicities, it is very useful to compare the multiplicity of M with the multiplicity of M/rM for some $r \in I$. Let $\operatorname{Ann}_r M$ denote the submodule of all elements $m \in M$ such that $rm = 0$.

PROPOSITION 2.73. *Let (R, m) be a local ring of dimension n, $I \subset R$ an m-primary ideal and M a finite R-module. Let $r \in I^d$ be an element and $\bar{r} \in \operatorname{gr}_I^d R$ its residue.*

Assume that $\operatorname{Supp} \operatorname{Ann}_r M$ has dimension $\leq n - 1$. Then

$$\operatorname{mult}_{(I/rI)}(M/rM) \geq d \cdot \operatorname{mult}_I(M),$$

and equality holds if $\operatorname{Supp} \operatorname{Ann}_{\bar{r}} \operatorname{gr}_I M$ has dimension $\leq n - 2$.

Proof. The multiplicity of M/rM is computed from the length of

$$(M/rM)/I^s(M/rM) = M/(I^s M + rM).$$

We have an exact sequence

$$0 \to K_{s-d} \to M/I^{s-d}M \xrightarrow{r} M/I^s M \to M/(I^s M + rM) \to 0,$$

where K_{s-d} is the kernel of multiplication by r. In general we know very little about K_{s-d}, but we at least have an inequality

$$HS_{(I/rI)}(M/rM, s) \geq HS_I(M, s) - HS_I(M, s - d),$$

proving $\operatorname{mult}_{(I/rI)}(M/rM) \geq d \cdot \operatorname{mult}_I(M)$.

If $\operatorname{Supp} \operatorname{Ann}_{\bar{r}} \operatorname{gr}_I M$ has dimension $\leq n - 2$, we can estimate the size of K_{s-d} from its filtered pieces for $0 \leq t \leq s - d$ since

$$(K_{s-d} \cap I^t M)/(K_{s-d} \cap I^{t+1} M) \subset (\operatorname{Ann}_{\bar{r}} \operatorname{gr}_I M)_t,$$

and so length $K_{s-d} \leq C \cdot (s-d)^{n-2}$. Thus we get that

$$HS_{(I/rI)}(M/rM, s) - HS_I(M, s) + HS_I(M, s-d) = O(s^{n-2}),$$

which gives the equality $\text{mult}_{(I/rI)}(M/rM) = d \cdot \text{mult}_I(M)$.   □

2.74 (Inductive computation of multiplicty). Let (R, m) be a local ring of dimension d and $I = (r_1, \ldots, r_t)$ an m-primary ideal. Let M be a finite R-module.

For $M_d := M$, $\text{gr}_I M_d$ is a module over $\text{gr}_I R$, and the latter is a quotient of the polynomial ring $(R/I)[x_1, \ldots, x_t]$. Let F be the sheaf corresponding to $\text{gr}_I M_d$ on $\mathbb{P}_{R/I}^{t-1}$ and $Z = \text{Supp}\, F \subset \mathbb{P}_{R/m}^{t-1}$. Note that $\dim Z \leq d-1$.

If R/m is infinite, then we can choose a d-dimensional subspace $V \subset \sum (R/m)x_i$ such that the corresponding projection $Z \to \mathbb{P}(V)_{R/m}$ is finite.

Let $\sum c_i x_i \in V$ be a general element, and set $r_d' = \sum c_i r_i \in R$. (One has to be somewhat careful here with writing $\sum c_i r_i$. This is well defined if R is an R/m-algebra, which is the main case that we are interested in. In general we have to choose a lifting of $c_i \in R/m$ to $c_i^* \in R$ and work with $\sum c_i^* r_i$. This makes no difference for our purposes.) Then $\bar{r}_d' = \sum c_i \bar{r}_i \in \text{gr}_I^1 R$ is not contained in any of the associated primes of $\text{gr}_I M$ with the possible exception of the maximal ideal of $\text{gr}_I R$. We get from (2.73) that

$$\text{mult}_I(M_d) = \text{mult}_{I/(r_d')}(M_d/r_d' M_d). \qquad (2.74.1)$$

Set $M_{d-1} := M_d/r_d' M_d$ and repeat the procedure. This way we get modules $M_d := M, M_{d-1}, \ldots, M_1$ such that

$$\text{mult}_I(M) = \text{mult}_{I/(r_{j+1}', \ldots, r_d')}(M_j) \quad \text{for } j \geq 1.$$

When we reach M_1, the maximal ideal corresponds to a codimension 1 point, and thus we cannot apply (2.73).

If $M = R$ and I is a maximal ideal, we recover one of the original heuristic definitions of multiplicities.

COROLLARY 2.75. *Let X be a scheme of dimension $d \geq 2$ over an infinite field k and $x \in X$ a closed point. Let $H_1, \ldots, H_{d-1}$ be general hyperplanes through x. Then $\text{mult}_x X = \text{mult}_x(X \cap H_1 \cap \cdots \cap H_{d-1})$.*   □

To continue, we have to take $M_1' := M_1/(0\text{-dimensional torsion})$, and then we can also get $M_0' := M_1'/r_1 M_1'$ with

$$\text{mult}_I M = \text{length}\, M_0' = \text{length}\, M_1'/r_1 M_1' \leq \text{length}\, M_1/r_1 M_1.$$

Let now J be the ideal generated by $V \subset \sum (R/m)r_i$, and consider $\text{mult}_J M$. Here $\text{gr}_J M_d$ gives a sheaf F_J on $\mathbb{P}(V)_{R/J}$ and $\text{Supp}\, F_J = \mathbb{P}(V)_{R/m}$. Thus we can assume that our choice of r_d' is such that it is not contained in any of the embedded primes of $\text{gr}_J M$ with the possible exception of the maximal ideal of $\text{gr}_J R$. With this choice of the r_i', the whole chain of modules $M_d := M, M_{d-1}, \ldots, M_1, M_1', M_0'$ is the same for J as for I.

Since M_0' is Artinian, its multiplicity does not depend on the ideal by (2.70.3). Thus we obtain the following.

COROLLARY 2.76. *Let (R, m) be a local ring of dimension d, $I = (r_1, \ldots, r_t)$ an m-primary ideal and M a finite R-module. Assume that R/m is infinite, and let $r_i' : i = 1, \ldots, d$ be general linear combinations of the r_i. Then*

$$\mathrm{mult}_I \, M = \mathrm{mult}_{(r_1', \ldots, r_d')} \, M \leq \mathrm{length} \, M/(r_1', \ldots, r_d')M. \quad \square$$

This implies another of the heuristic definitions of multiplicities.

COROLLARY 2.77. *Let $X \subset \mathbb{A}^n$ be a scheme of dimension $d \geq 1$ over an infinite field k and $x \in X$ a closed point. Let $\pi : X \to \mathbb{A}^{d+1}$ be a general projection. Then $\mathrm{mult}_x X = \mathrm{mult}_{\pi(x)} \pi(X)$.*

Proof. By (2.76) we can choose linear forms $h_1, \ldots, h_d$ vanishing at x such that $\mathrm{mult}_x X = \mathrm{mult}_{(h_1, \ldots, h_d)} \mathcal{O}_{x, X}$. Choose one more h_{d+1} such that the projection to $\mathbb{A}^{d+1}$ given by the $h_1, \ldots, h_{d+1}$ is birational and that the preimage of $\pi(x)$ is x. Then

$$\mathrm{mult}_{\pi(x)} \pi(X) = \mathrm{mult}_{(h_1, \ldots, h_{d+1})} \mathcal{O}_{\pi(x), \pi(X)} = \mathrm{mult}_{(h_1, \ldots, h_{d+1})} \mathcal{O}_{x, X},$$

the latter by (2.72.1). Furthermore, we have the trivial inequalities

$$\mathrm{mult}_{(h_1, \ldots, h_d)} \mathcal{O}_{x, X} \leq \mathrm{mult}_{(h_1, \ldots, h_{d+1})} \mathcal{O}_{x, X} \leq \mathrm{mult}_x X,$$

which are equalities by our choice of the $h_1, \ldots, h_d$. $\quad \square$

PROPOSITION 2.78. *Let X be a scheme and $x \in X$ a closed point. Then $\mathrm{mult}_{x'} B_x X \leq \mathrm{mult}_x X$ for every $x' \in E \subset B_x X$.*

Proof. Let $E = \mathrm{Proj}(\mathrm{gr}_m \mathcal{O}_{x, X}) \subset B_x X$ be the exceptional divisor. The multiplicity $\mathrm{mult}_x X$ is the degree of E in its natural projective embedding, and thus $\mathrm{mult}_{x'} E \leq \deg E = \mathrm{mult}_x X$ for every $x' \in E$. Thus $\mathrm{mult}_{x'} B_x X \leq \mathrm{mult}_x X$ by (2.73). $\quad \square$

2.79 (Generalized Bézout theorem). Let $X \subset \mathbb{P}_k^n$ be a projective variety of dimension d over a field k and $H_1, \ldots, H_d \subset \mathbb{P}^n$ hypersurfaces such that $X \cap H_1 \cap \cdots \cap H_d$ is a finite set of points.

For any $x \in X$, take local defining equations $H_i = (f_{ip} = 0)$. If X is smooth, then (see, for instance, [Sha94, Sec.IV.1])

$$\deg X \cdot \prod \deg H_i = \sum_{p \in X} \dim_k \mathcal{O}_{p, X}/(f_{1p}, \ldots, f_{dp}). \qquad (2.79.1)$$

One of the difficulties of early intersection theory was that this formula fails for singular varieties.

EXAMPLE 2.80. Let $S = \mathbb{C}[x^2, xy, y^2, x^3, x^2y, xy^2, y^3]$, that is, $\mathbb{C}[x, y]$ without the linear polynomials, and set $X := \operatorname{Spec} S \subset \mathbb{A}^7(u_1, \ldots, u_7)$. Let $\pi : \mathbb{A}^7 \to \mathbb{A}^2$ be the projection to the (u_1, u_3)-coordinates. Thus the coordinate ring of the image $\mathbb{A}^2$ is identified with $\mathbb{C}[x^2, y^2]$.

The quotient field of S is $\mathbb{C}(x, y)$, and the field extension induced by the projection $\pi : S \to \mathbb{A}^2$ is $\mathbb{C}(x, y)/\mathbb{C}(x^2, y^2)$, which has degree 4. Correspondingly, if $c_1, c_3 \neq 0$, then the 5-plane $(u_1 - c_1 = u_3 - c_3 = 0)$ intersects Z transversally in four distinct points corresponding to $x = \pm\sqrt{c_1}, y = \pm\sqrt{c_3}$.

However, for $(u_1 = u_3 = 0)$, we get the local term
$$\dim \mathbb{C}[x^2, xy, y^2, x^3, x^2y, xy^2, y^3]/(x^2, y^2) = \dim\langle 1, xy, x^3, x^2y, xy^2, y^3\rangle = 6.$$

The correct answer is given by the Hilbert-Samuel multiplicities.

THEOREM 2.81 (General Bézout theorem). *Let $X \subset \mathbb{P}^n_k$ be a projective scheme of pure dimension d over a field k and $H_1, \ldots, H_d \subset \mathbb{P}^n$ hypersurfaces such that $X \cap H_1 \cap \cdots \cap H_d$ is a finite set of points. Then*
$$\deg X \cdot \prod \deg H_i = \sum_{p \in X} \operatorname{mult}_{(f_{1p}, \ldots, f_{dp})} \mathcal{O}_{p,X}.$$

Proof. Let R be the local ring of the vertex of the cone over X in $\mathbb{A}^{n+1}$. Let $H_i = (f_i = 0)$, and choose a linear form f_0 not vanishing at all points of $X \cap H_1 \cap \cdots \cap H_d$. Assume that $\deg f_0 \leq \cdots \leq \deg f_d$. We prove that both sides are equal to $\operatorname{mult}_{(f_0, \ldots, f_d)} R$.

Let f_i' be a general homogeneous element of $(f_0, \ldots, f_d)$ of the same degree as f_i. Then (2.73) can be applied inductively starting with f_d' to conclude that
$$\operatorname{mult}_{(f_0, \ldots, f_d)} R = \operatorname{mult}_{(f_0', \ldots, f_d')} R = \deg X \cdot \prod_i \deg f_i.$$

Next let f_i^* be a general homogeneous element of $(f_1, \ldots, f_d)$ of the same degree as f_i. Then (2.73) can again be applied inductively, except at $R/(f_2^*, \ldots, f_d^*)$ we have to take the quotient by the torsion submodule to get M_2, and then we get $M_1 = M_2/f_1^* M_2$.

The 0-dimensional torsion in M_1 is unclear, but the 1-dimensional part $M_1' := M_1/(0\text{-dimensional torsion})$ is the homogeneous coordinate ring of the lines in $\mathbb{A}^{n+1}$ over the intersection points $X \cap H_1 \cap \cdots \cap H_d$. The line over p appears with multiplicity $\operatorname{mult}_{(f_{1p}, \ldots, f_{dp})} \mathcal{O}_{p,X}$ since we can use $f_{ip} := f_i^*/f_0^{\deg f_i}$ as local equations at p to go through the same computations locally in each $\mathcal{O}_{p,X}$. At the end we get that
$$\operatorname{mult}_{(f_0, \ldots, f_d)} R = \dim_k(M_1'/f_0 M_1') = \sum_{p \in X} \operatorname{mult}_{(f_{1p}, \ldots, f_{dp})} \mathcal{O}_{p,X}. \qquad \square$$

CHAPTER 3

Strong Resolution in Characteristic Zero

The most influential paper on resolution of singularities is Hironaka's magnum opus [Hir64]. Its starting point is a profound shift in emphasis from resolving singularities of varieties to resolving "singularities of ideal sheaves." Ideal sheaves of smooth or simple normal crossing divisors are the simplest ones. Locally, in a suitable coordinate system, these ideal sheaves are generated by a single monomial. The aim is to transform an arbitrary ideal sheaf into such a "locally monomial" one by a sequence of blow-ups. Ideal sheaves are much more flexible than varieties, and this opens up new ways of running induction.

Since then, resolution of singularities emerged as a very unusual subject whose main object has been a deeper understanding of the proof, rather than the search for new theorems. A better grasp of the proof leads to improved theorems, with the ultimate aim of extending the method to positive characteristic. Two seemingly contradictory aspects make it very interesting to study and develop Hironaka's approach.

First, the method is very robust, in that many variants of the proof work. One can even change basic definitions and be rather confident that the other parts can be modified to fit.

Second, the complexity of the proof is very sensitive to details. Small changes in definitions and presentation may result in major simplifications.

This duality also makes it difficult to write a reasonable historical presentation and to correctly appreciate the contributions of various researchers. Each step ahead can be viewed as small or large, depending on whether we focus on the change in the ideas or on their effect. In some sense, all the results of the past forty years have their seeds in [Hir64], nevertheless, the improvement in the methods has been enormous. Thus, instead of historical notes, here is a list of the most important contributions to the development of the Hironaka method, more or less in historical order: Hironaka [Hir64, Hir77]; Giraud [Gir74]; Villamayor [Vil89, Vil92, Vil96] with his coworkers Bravo [BV01] and Encinas [EV98, EV03]; Bierstone and Milman [BM89, BM91, BM97, BM03]; Encinas and Hauser [EH02] and Włodarczyk [Wło05]. The following proof relies mostly on the works of Villamayor and Włodarczyk.

The methods of Bierstone and Milman and of Encinas and Hauser differ from ours (and from each other) in key technical aspects, though the actual resolution procedures end up very similar.

I have also benefited from the surveys and books [Gir95, Lip75, AHV77, CGO84, HLOQ00, Hau03, Cut04].

Abhyankar's book [Abh66] shows some of the additional formidable difficulties that appear in positive characteristic.

A very elegant approach to resolution following de Jong's results on alterations [dJ96] is developed in the papers [BP96, AdJ97, AW97]. This method produces a resolution as in (3.2), which is however neither strong (3.3) nor functorial (3.4). The version given in [Par99] is especially simple.

Another feature of the study of resolutions is that everyone seems to use different terminology, so I also felt free to introduce my own.

It is very instructive to compare the current methods with Hironaka's "idealistic" paper [Hir77]. The main theme is that resolution becomes simpler if we do not try to control the process very tightly, as illustrated by the following three examples.

(1) The original method of [Hir64] worked with the Hilbert-Samuel function of an ideal sheaf at a point. It was gradually realized that the process simplifies if one considers only the vanishing order of an ideal sheaf—a much cruder invariant.

(2) Two ideals I and J belong to the same *idealistic exponent* if they behave "similarly" with respect to any birational map g. (That is, g^*I and g^*J agree at the generic point of every divisor for every g.) Now we see that it is easier to work with an equivalence relation that requires the "similar" behavior only with respect to some birational maps (namely, composites of smooth blow-ups along subvarieties, where the vanishing order is maximal).

(3) The concept of a *distinguished presentation* attempts to pick a local coordinate system that is optimally adjusted to the resolution of a variety or ideal sheaf. A key result of Włodarczyk [Wło05] says that for a suitably modified ideal, all reasonable choices are equivalent, and thus we do not have to be very careful. Local coordinate systems are not needed at all.

The arguments given here differ from their predecessors in two additional aspects. The first of these is a matter of choice, but the second one makes the structure of the proof patent.

(4) The inductive proof gives resolutions only locally, and patching the local resolutions has been quite difficult. The best way would be to define an invariant on points of varieties $\mathrm{inv}(x, X)$ with values in an ordered set such that

(i) $x \mapsto \mathrm{inv}(x, X)$ is an upper semi continuous function, and

(ii) at each step of the resolution we blow up the locus where the invariant is maximal.

With some modification, this is accomplished in [Vil89, BM97, EH02]. All known invariants are, however, rather complicated. Włodarczyk suggested in [Wło05] that with his methods it should not be necessary to define such an invariant. We show that, by a slight change in the definitions, the resolution algorithm automatically globalizes, obviating the need for the invariant.

(5) Traditionally, the results of Sections 9–12 constituted one intertwined package, which had to be carried through the whole induction together. The introduction of the notions of *D-balanced* and *MC-invariant* ideal sheaves makes it possible to disentangle these to obtain four independent parts.

3.1. What is a good resolution algorithm?

Before we consider the resolution of singularities in general, it is worthwhile to contemplate what the properties of a good resolution algorithm should be.

Here I concentrate on the case of resolving singularities of varieties only. In practice, one may want to keep track and improve additional objects, for instance, subvarieties or sheaves as well, but for now these variants would only obscure the general picture.

3.1 (Weakest resolution). *Given a variety X, find a projective variety X' such that X' is smooth and birational to X.*

This is what the Albanese method gives for curves and surfaces. In these cases one can then use this variant to get better resolutions, so we do not lose anything at the end. These stronger forms are, however, not automatic, and it is not at all clear that such a "weakest resolution" would be powerful enough in higher dimensions.

(Note that even if X is not proper, we have to insist on X' being proper; otherwise, one could take the open subset of smooth points of X for X'.)

In practice it is useful, sometimes crucial, to have additional properties.

3.2 (Resolution). *Given a variety X, find a variety X' and a projective morphism $f : X' \to X$ such that X' is smooth and f is birational.*

This is the usual definition of resolution of singularities.

For many applications this is all one needs, but there are plenty of situations when additional properties would be very useful. Here are some of these.

3.2.1 (Singularity theory). Let us start with an isolated singularity $x \in X$. One frequently would like to study it by taking a resolution $f : X' \to X$ and connecting the properties of $x \in X$ with properties of the exceptional divisor $E = \mathrm{Ex}(f)$. Here everything works best if E is projective, that is, when $E = f^{-1}(x)$.

It is reasonable to hope that we can achieve this. Indeed, by assumption, $X \setminus \{x\}$ is smooth, so it should be possible to resolve without changing $X \setminus \{x\}$.

3.2.2 (Open varieties). It is natural to study a noncompact variety X^0 via a compactification $X \supset X^0$. Even if X^0 is smooth, the compactifications that are easy to obtain are usually singular. Then one would like to resolve the singularities of X and get a smooth compactification X'. If we take any resolution $f : X' \to X$, the embedding $X^0 \hookrightarrow X$ does not lift to an embedding $X^0 \hookrightarrow X'$. Thus we would like to find a resolution $f : X' \to X$ such that f is an isomorphism over X^0.

In both of the above examples, we would like the exceptional set E or the boundary $X' \setminus X^0$ to be "simple." Ideally we would like them to be smooth, but this is rarely possible. The next best situation is when E or $X' \setminus X^0$ are simple normal crossing divisors.

These considerations lead to the following variant.

3.3 (Strong resolution). *Given a variety X, find a variety X' and a projective morphism $f : X' \to X$ such that we have the following:*

(1) *X' is smooth and f is birational,*
(2) *$f : f^{-1}(X^{ns}) \to X^{ns}$ is an isomorphism, and*
(3) *$f^{-1}(\operatorname{Sing} X)$ is a divisor with simple normal crossings.*

Here $\operatorname{Sing} X$ denotes the set of *singular points* of X and $X^{ns} := X \setminus \operatorname{Sing} X$ the set of *smooth points*.

Strong resolution seems to be the variant that is most frequently used in applications, but sometimes other versions are needed. For instance, one might need condition (3.3.3) scheme theoretically.

A more important question arises when one has several varieties X_i to work with simultaneously. In this case we may need to know that certain morphisms $\phi_{ij} : X_i \to X_j$ lift to the resolutions $\phi'_{ij} : X'_i \to X'_j$.

It would be nice to have this for all morphisms, which would give a "resolution functor" from the category of all varieties and morphisms to the category of smooth varieties. This is, however, impossible.

Example 3.3.4. Let $S := (uv - w^2 = 0) \subset \mathbb{A}^3$ be the quadric cone, and consider the morphism

$$\phi : \mathbb{A}^2_{x,y} \to S \quad \text{given by} \quad (x, y) \mapsto (x^2, y^2, xy).$$

The only sensible resolution of $\mathbb{A}^2$ is itself, and any resolution of S dominates the minimal resolution $S' \to S$ obtained by blowing up the origin.

The morphism ϕ lifts to a rational map $\phi' : \mathbb{A}^2 \dashrightarrow S'$, but ϕ' is not a morphism.

It seems that the best one can hope for is that the resolution commutes with smooth morphisms.

3.4 (Functorial resolution). *For every variety X find a resolution f_X : $X' \to X$ that is functorial with respect to smooth morphisms. That is, any smooth morphism $\phi : X \to Y$ lifts to a smooth morphism $\phi' : X' \to Y'$, which gives a fiber product square*

$$
\begin{array}{ccc}
X' & \xrightarrow{\phi'} & Y' \\
f_X \downarrow & \square & \downarrow f_Y \\
X & \xrightarrow{\phi} & Y.
\end{array}
$$

Note that if ϕ' exists, it is unique, and so we indeed get a functor from the category of all varieties and smooth morphisms to the category of smooth varieties and smooth morphisms.

This is quite a strong property with many useful implications.

3.4.1 (Group actions). Functoriality of resolutions implies that any group action on X lifts to X'. For discrete groups this is just functoriality plus the observation that the only lifting of the identity map on X is the identity map of X'. For an algebraic group G a few more steps are needed; see (3.9.1).

3.4.2 (Localization). Let $f_X : X' \to X$ be a functorial resolution. The embedding of any open subset $U \hookrightarrow X$ is smooth, and so the functorial resolution of U is the restriction of the functorial resolution of X. That is,

$$
\left(f_U : U' \to U \right) \cong \left(f_X |_{f_X^{-1}(U)} : f_X^{-1}(U) \to U \right).
$$

Equivalently, a functorial resolution is Zariski local. More generally, a functorial resolution is étale local since étale morphisms are smooth.

Conversely, we show in (3.9.2) that any resolution that is functorial with respect to étale morphisms is also functorial with respect to smooth morphisms.

Since any resolution $f : X' \to X$ is birational, it is an isomorphism over some smooth points of X. Any two smooth points of X are étale equivalent, and thus a resolution that is functorial with respect to étale morphisms is an isomorphism over smooth points. Thus any functorial resolution satisfies (3.3.2).

3.4.3 (Formal localization). Any sensible étale local construction in algebraic geometry is also formal local. In our case this means that the behavior of the resolution $f_X : X' \to X$ near a point $x \in X$ should depend only on the completion $\widehat{\mathcal{O}}_{x,X}$. (Technically speaking, $\operatorname{Spec}\widehat{\mathcal{O}}_{x,X}$ is not a variety and the map $\operatorname{Spec}\widehat{\mathcal{O}}_{x,X} \to \operatorname{Spec}\mathcal{O}_{x,X}$ is only formally smooth, so this is a stronger condition than functoriality.)

3.4.4 (Resolution of products). It may appear surprising, but a strong and functorial resolution should *not* commute with products.

For instance, consider the quadric cone $0 \in S = (x^2 + y^2 + z^2 = 0) \subset \mathbb{A}^3$. This is resolved by blowing up the origin $f : S' \to S$ with exceptional curve

$C \cong \mathbb{P}^1$. On the other hand,

$$f \times f : S' \times S' \to S \times S$$

cannot be the outcome of an étale local strong resolution. The singular locus of $S \times S$ has two components, $Z_1 = \{0\} \times S$ and $Z_2 = S \times \{0\}$, and correspondingly, the exceptional divisor has two components, $E_1 = C \times S'$ and $E_2 = S' \times C$, which intersect along $C \times C$.

If we work étale locally at $(0, 0)$, we cannot tell whether the two branches of the singular locus $Z_1 \cup Z_2$ are on different irreducible components of $\operatorname{Sing} S$ or on one non-normal irreducible component. Correspondingly, the germs of E_1 and E_2 could be on the same irreducible exceptional divisor, and on a strong resolution self-intersections of exceptional divisors are not allowed.

So far we concentrated on the end result $f_X : X' \to X$ of the resolution. Next we look at some properties of the resolution algorithm itself.

3.5 (Resolution by blowing up smooth centers). *For every variety X find a resolution $f_X : X' \to X$ such that f_X is a composite of morphisms*

$$f_X : X' = X_n \xrightarrow{p_{n-1}} X_{n-1} \xrightarrow{p_{n-2}} \cdots \xrightarrow{p_1} X_1 \xrightarrow{p_0} X_0 = X,$$

where each $p_i : X_{i+1} \to X_i$ is obtained by blowing up a smooth subvariety $Z_i \subset X_i$.

If we want $f_X : X' \to X$ to be a strong resolution, then the condition $Z_i \subset \operatorname{Sing} X_i$ may also be required, though we need only that $p_0 \cdots p_{i-1}(Z_i) \subset \operatorname{Sing} X$.

Let us note first that in low dimensions some of the best resolution algorithms do not have this property.

(1) The quickest way to resolve a curve is to normalize it. The normalization usually cannot be obtained by blowing up points (though it is a composite of blow-ups of points).

(2) A normal surface can be resolved by repeating the procedure: "blow up the singular points and normalize" [Zar39].

(3) A toric variety is best resolved by toric blow-ups. These are rarely given by blow-ups of subvarieties (cf. [Ful93, 2.6]).

(4) Many of the best-studied singularities are easier to resolve by doing a weighted blow-up first.

(5) The theory of Nash blow-ups offers a—so far mostly hypothetical—approach to resolution that does not rely on blowing up smooth centers; cf. [Hir83].

On the positive side, resolution by blowing up smooth centers has the great advantage that we do not mess up what is already nice. For instance, if we want to resolve X and $Y \supset X$ is a smooth variety containing X, then a resolution by blowing up smooth centers automatically carries along the smooth variety. Thus we get a sequence of smooth varieties Y_i fitting in a diagram

$$
\begin{array}{ccccccc}
X_n & \xrightarrow{p_{n-1}} & X_{n-1} & \cdots & X_1 & \xrightarrow{p_0} & X_0 = X \\
\downarrow & & \downarrow & & \downarrow & & \downarrow \\
Y_n & \xrightarrow{q_{n-1}} & Y_{n-1} & \cdots & Y_1 & \xrightarrow{q_0} & Y_0 = Y,
\end{array}
$$

where the vertical arrows are closed embeddings.

Once we settle on resolution by successive blowing ups, the main question is how to find the centers that we need to blow up. From the algorithmic point of view, the best outcome would be the following.

3.6 (Iterative resolution, one blow-up at a time). *For any variety X, identify a subvariety $W(X) \subset X$ consisting of the "worst" singularities. Set $R(X) := B_{W(X)}X$ and $R^m(X) := R(R^{m-1}(X))$ for $m \geq 2$. Then we get resolution by iterating this procedure. That is, $R^m(X)$ is smooth for $m \gg 1$.*

Such an algorithm exists for curves with $W(X) = \operatorname{Sing} X$.

The situation is not so simple in higher dimensions.

Example 3.6.1. Consider the pinch point, or Whitney umbrella, $S := (x^2 - y^2 z = 0) \subset \mathbb{A}^3$. S is singular along the line $(x = y = 0)$. It has a normal crossing point if $z \neq 0$ but a more complicated singularity at $(0, 0, 0)$.

If we blow up the "worst" singular point $(0, 0, 0)$ of the surface S, then in the chart with coordinates $x_1 = x/z, y_1 = y/z, z_1 = z$ we get the birational transform $S_1 = (x_1^2 - y_1^2 z_1 = 0)$. This is isomorphic to the original surface.

Thus we conclude that one cannot resolve surfaces by blowing up the "worst" singular point all the time.

We can, however, resolve the pinch point by blowing up the whole singular line. In this case, using the multiplicity (which is a rough invariant) gives the right blow-up, whereas distinguishing the pinch point from a normal crossing point (using some finer invariants) gives the wrong blow-up. The message is that we should not look at the singularities too carefully.

The situation gets even worse for normal 3-folds.

Example 3.6.2. Consider the 3-fold

$$
X := (x^2 + y^2 + z^m t^m = 0) \subset \mathbb{A}^4.
$$

The singular locus is the union of the two lines

$$
L_1 := (x = y = z = 0) \quad \text{and} \quad L_2 := (x = y = t = 0).
$$

There are two reasons why no sensible resolution procedure should start by blowing up either of the lines.

(i) The two lines are interchanged by the involution $\tau : (x, y, z, t) \mapsto (x, y, t, z)$, and thus they should be blown up in a τ-invariant way.

  (ii) An étale local resolution procedure cannot tell if $L_1 \cup L_2$ is a union
       of two lines or just two local branches of an irreducible curve. Thus
       picking one branch does not make sense globally.

Therefore, we must start by blowing up the intersection point $(0,0,0,0)$ (or
resort to blowing up a singular subscheme).
  Computing the t-chart $x = x_1 t_1, y = y_1 t_1, z = z_1 t_1, t = t_1$, we get

$$X_{1,t} = (x_1^2 + y_1^2 + z_1^m t_1^{2m-2} = 0)$$

and similarly in the z-chart. Thus on $B_0 X$ the singular locus consists of
three lines: L_1', L_2' and an exceptional line E.
  For $m = 2$ we are thus back to the original situation, and for $m \geq 3$
we made the singularities worse by blowing up. In the $m = 2$ case there is
nothing else one can do, and we get our first negative result.

Claim 3.6.3. There is no iterative resolution algorithm that works one
smooth blow-up at a time.

  The way out is to notice that our two objections (3.6.2.i–ii) to first
blowing up one of the lines L_1 or L_2 are not so strong when applied to
the three lines L_1, L_2 and E on the blow-up $B_0 X$. Indeed, we know that
the new exceptional line E is isomorphic to $\mathbb{CP}^1$, and it is invariant under
every automorphism lifted from X. Thus we can safely blow up $E \subset B_0 X$
without the risk of running into problems with étale localization. (A key
point is that we want to ensure that the process is étale local only on X,
not on all the intermediate varieties.) In the $m = 2$ case we can then blow
up the birational transforms of the two lines L_1 and L_2 simultaneously, to
achieve resolution. (Additional steps are needed for $m \geq 3$.)
  In general, we have to ensure that the resolution process has some
"memory." That is, at each step the procedure is allowed to use information about the previous blow-ups. For instance, it could keep track of the
exceptional divisors that were created by earlier blow-ups of the resolution
and in which order they were created.

  3.7 (Other considerations). There are several other ways to judge how
good a resolution algorithm is.
  3.7.1 (Elementary methods). A good resolution method should be part
of "elementary" algebraic geometry. Both Newton's method of rotating
rulers and the Albanese projection method pass this criterion. On the other
hand, several of the methods for surfaces rely on more advanced machinery.
  3.7.2 (Computability). In concrete cases, one may wish to explicitly
determine resolutions by hand or by a computer. As far as I can tell,
the existing methods do rather poorly even on the simplest singularities.
In a more theoretical direction, one can ask for the worst case or average

complexity of the algorithms. See [BS00b, BS00a, FKP05] for computer implementations.

3.8. Our resolution is strong and functorial with respect to smooth morphisms, but it is very far from being iterative if we want to work one blow-up at a time. Instead, at each step we specify a long sequence of blow-ups to be performed.

We shift our emphasis from resolution of singularities to principalization of ideal sheaves. While principalization is achieved by a sequence of smooth blow-ups, the resolution of singularities may involve blow-ups of singular centers. Furthermore, at some stage we may blow up a subvariety $Z_i \subset X_i$ along which the variety X_i is smooth. This only happens for subvarieties that sit over the original singular locus, so at the end we still get a strong resolution.

The computability of the algorithm has not been investigated much, but the early indications are not promising. One issue is that starting with, say, a hypersurface $(f = 0) \subset \mathbb{A}^n$ of multiplicity m the first step is to replace the ideal (f) with another ideal $W(f)$, which has more than e^{mn} generators, each of multiplicity at least e^m; see (3.54.3). Then we reduce to a resolution problem in $(n-1)$-dimensions, and at the next reduction step we again may have an exponential increase of the multiplicity and the number of generators. For any reasonable computer implementation, some shortcuts are essential.

ASIDE 3.9. Here we prove the two claims made in (3.4). These are not used in the rest of the chapter.

Proposition 3.9.1. The action of an algebraic group G on a scheme X lifts to an action of G on its functorial resolution X'.

Proof. The action of an algebraic group G on a variety X is given by a smooth morphism $m : G \times X \to X$. By functoriality, the resolution $(G \times X)'$ of $G \times X$ is given by the pull-back of X' via m, that is, by $f_X^*(m) : (G \times X)' \to X'$.

On the other hand, the second projection $\pi_2 : G \times X \to X$ is also smooth, and so $(G \times X)' = G \times X'$. Thus we get a commutative diagram

$$
\begin{array}{ccccc}
m' : G \times X' & \cong & (G \times X)' & \xrightarrow{f_X^*(m)} & X' \\
\downarrow id_G \times f_X & & & & \downarrow f_X \\
G \times X & = & G \times X & \xrightarrow{m} & X.
\end{array}
$$

We claim that the composite in the top row $m' : G \times X' \to X'$ defines a group action. This means that the following diagram is commutative, where $m_G : G \times G \to G$ is the group multiplication:

$$
\begin{array}{ccc}
G \times G \times X' & \xrightarrow{id_G \times m'} & G \times X' \\
m_G \times id_{X'} \downarrow & & \downarrow m' \\
G \times X' & \xrightarrow{m'} & X'.
\end{array}
$$

Since $m : G \times X \to X$ defines a group action, we know that the diagram is commutative over a dense open set. Since all schemes in the diagram are separated and reduced, this implies commutativity.   □

Proposition 3.9.2. Any resolution that is functorial with respect to étale morphisms is also functorial with respect to smooth morphisms.

Proof. As we noted in (3.4.2), a resolution that is functorial with respect to étale morphisms is an isomorphism over smooth points.

Étale locally, a smooth morphism is a direct product, and so it is sufficient to prove that $(X \times A)' \cong X' \times A$ for any abelian variety A. Such an isomorphism is unique; thus it is enough to prove existence for X proper.

Since $(X \times A)'$ is proper, the connected component of its automorphism group is an algebraic group G (see, for instance, [Kol96, I.1.10]). Let $G_1 \subset G$ denote the subgroup whose elements commute with the projection $\pi : (X \times A)' \to X$.

Let $Z \subset X^{ns}$ be a finite subset. Then $\pi^{-1}(Z) \cong Z \times A$, and the action of A on itself gives a subgroup $j_Z : A \hookrightarrow \mathrm{Aut}(\pi^{-1}(Z))$. There is a natural restriction map $\sigma_Z : G_1 \to \mathrm{Aut}(\pi^{-1}(Z))$; set $G_Z := \sigma_Z^{-1}(j_Z A)$.

As we increase Z, the subgroups G_Z form a decreasing sequence, which eventually stabilizes at a subgroup $G_X \subset G$ such that for every finite set $Z \subset X^{ns}$ the action of G_X on $\pi^{-1}(Z)$ is through the action of A on itself. This gives an injective homomorphism of algebraic groups $G_Z \hookrightarrow A$.

On the other hand, A acts on $X \times A$ by isomorphisms, and by assumption this action lifts to an action of the *discrete* group A on $(X \times A)'$. Thus the injection $G_Z \hookrightarrow A$ has a set-theoretic inverse, so it is an isomorphism of algebraic groups.   □

3.2. Examples of resolutions

We start the study of resolutions with some examples. First, we describe how the resolution method deals with two particular surface singularities $S \subset \mathbb{A}^3$. While these are relatively simple cases, they allow us to isolate six problems facing the method. Four of these we solve later, and we can live with the other two.

Then we see how the problems can be tackled for Weierstrass polynomials and what this solution tells us about the general case. For curves and surfaces, this method was already used in Sections 1.10 and 2.7.

KEY IDEA 3.10. We look at the trace of $S \subset \mathbb{A}^3$ on a suitable smooth surface $H \subset \mathbb{A}^3$ and reconstruct the whole resolution of S from $S \cap H$.

More precisely, starting with a surface singularity $0 \in S \subset \mathbb{A}^3$ of multiplicity m, we will be guided by $S \cap H$ until the multiplicity of the birational transform (2.1) of S drops below m. Then we need to repeat the method to achieve further multiplicity reduction.

EXAMPLE 3.11 (Resolving $S := (x^2 + y^3 - z^6 = 0) \subset \mathbb{A}^3$). (We already know from (2.4) that the minimal resolution has a single exceptional curve $E \cong (x^2 z + y^3 - z^3 = 0) \subset \mathbb{P}^2$ and it has self-intersection $(E^2) = -1$ but let us forget it for now.)

Set $H := (x = 0) \subset \mathbb{A}^3$, and work with $S \cap H$.

Step 1. Although the trace $S \cap H = (y^3 - z^6 = 0) \subset \mathbb{A}^2$ has multiplicity 3, we came from a multiplicity 2 situation, and we blow up until the multiplicity drops below 2.

Here it takes two blow-ups to achieve this. The crucial local charts and equations are

$$x^2 + y^3 - z^6 = 0,$$
$$x_1^2 + (y_1^3 - z_1^3)z_1 = 0, \qquad x_1 = x/z, y_1 = y/z, z_1 = z,$$
$$x_2^2 + (y_2^3 - 1)z_2^2 = 0, \qquad x_2 = x_1/z_1, y_2 = y_1/z_1, z_2 = z_1.$$

At this stage the trace of the dual graph (2.6) of the birational transform of S on the birational transform of H is the following, where the numbers indicate the multiplicity (and not minus the self-intersection number as usual) and $\bullet$ indicates the birational transform of the original curve $S \cap H$:

Step 2. The birational transform of $S \cap H$ intersects some of the new exceptional curves that appear with positive coefficient. We blow up until these intersections are removed.

In our case each intersection point needs to be blown up twice. At this stage the trace of the birational transform of S on the birational transform of H looks like

$$
\begin{array}{ccccccc}
 & & 1 & - & 0 & - & \bullet \\
 & \nearrow & & & & & \\
1 & - & 2 & - & 1 & - & 0 & - & \bullet \\
 & \searrow & & & & & \\
 & & 1 & - & 0 & - & \bullet
\end{array}
$$

where multiplicity 0 indicates that the curve is no longer contained in the birational transform of H (so strictly speaking, we should not draw it at all).

Step 3. The trace now has multiplicity < 2 along the birational transform of $S \cap H$, but it still has some points of multiplicity ≥ 2. We remove these by blowing up the exceptional curves with multiplicity ≥ 2.

In our case there is only one such curve. After blowing it up, we get the final picture

$$
\begin{array}{ccccc}
& & 1 & - \ 0 \ - & \bullet \\
& \diagup & & & \\
1 - \boxed{0} & - & 1 & - \ 0 \ - & \bullet \\
& \diagdown & & & \\
& & 1 & - \ 0 \ - & \bullet
\end{array}
$$

where the boxed curve is elliptic.

More details of the resolution method appear in the following example.

EXAMPLE 3.12 (Resolving $S := (x^3 + (y^2 - z^6)^2 + z^{21}) = 0) \subset \mathbb{A}^3$). As before, we look at the trace of S on the plane $H := (x = 0)$ and reconstruct the whole resolution of S from $S \cap H$.

Step 1. Although the trace $S \cap H = ((y^2 - z^6)^2 + z^{21} = 0) \subset \mathbb{A}^2$ has multiplicity 4, we came from a multiplicity 3 situation, and we blow up until the multiplicity drops below 3.

Here it takes three blow-ups to achieve this. The crucial local charts and equations are

$$
\begin{aligned}
x^3 + (y^2 - z^6)^2 + z^{21} &= 0, \\
x_1^3 + z_1(y_1^2 - z_1^4)^2 + z_1^{18} &= 0, & x_1 = x/z, y_1 = y/z, z_1 = z, \\
x_2^3 + z_2^2(y_2^2 - z_2^2)^2 + z_2^{15} &= 0 & x_2 = x_1/z_1, y_2 = y_1/z_1, z_2 = z_1, \\
x_3^3 + z_3^3(y_3^2 - 1)^2 + z_3^{12} &= 0, & x_3 = x_2/z_2, y_3 = y_2/z_2, z_3 = z_2.
\end{aligned}
$$

The birational transform of $S \cap H$ has equation

$$
(y_3^2 - 1)^2 + z_3^9 = 0
$$

and has two higher cusps at $y_3 = \pm 1$ on the last exceptional curve. The trace of the birational transform of S on the birational transform of H looks like

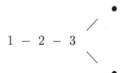

$$
\begin{array}{ccc}
& & \bullet \\
& & \diagup \\
1 - 2 - 3 & & \\
& & \diagdown \\
& & \bullet
\end{array}
$$

(As before, the numbers indicate the multiplicity, and $\bullet$ indicates the birational transform of the original curve $S \cap H$. Also note that here the curves marked • have multiplicity 2 at their intersection point with the curve marked 3.)

Step 2. The birational transform of $S \cap H$ intersects some of the new exceptional curves that appear with positive coefficient. We blow up until these intersections are removed.

In our case each intersection point needs to be blown up three times, and we get the following picture:

$$
\begin{array}{ccccccc}
& & 2 & - & 1 & - & 0 & - & \bullet \\
& \diagup & & & & & & \\
1 & - & 2 & - & 3 & & & \\
& & & \diagdown & & & & \\
& & & & 2 & - & 1 & - & 0 & - & \bullet
\end{array}
$$

Step 3. The trace now has multiplicity < 3 along the birational transform of $S \cap H$, but it still has some points of multiplicity ≥ 3. There is one exceptional curve with multiplicity ≥ 3; we blow that up. This drops its coefficient from 3 to 0. There are four more points of multiplicity 3, where a curve with multiplicity 2 intersects a curve with multiplicity 1. After blowing these up we get the final picture

$$
\begin{array}{ccccccccc}
& & 2 & - & 0 & - & 1 & - & 0 & - & \bullet \\
& \diagup & & & & & & & \\
1 & - & 0 & - & 2 & - & 0 & & & \\
& & & \diagdown & & & & & \\
& & & & 2 & - & 0 & - & 1 & - & 0 & - & \bullet
\end{array}
$$

3.13 (Problems with the method). There are at least six different problems with the method. Some are clearly visible from the examples, while some are hidden by the presentation.

Problem 3.13.1. In (3.11) we end up with eight exceptional curves, when we need only one to resolve S; see (2.4). In general, for many surfaces the method gives a resolution that is much bigger than the minimal one. However, in higher dimensions there is no minimal resolution, and it is not clear how to measure the "wastefulness" of a resolution.

We will not be able to deal with this issue.

Problem 3.13.2. The resolution problem for surfaces in $\mathbb{A}^3$ was reduced not to the resolution problem for curves in $\mathbb{A}^2$ but to a related problem that also takes into account exceptional curves and their multiplicities in some way.

We have to set up a somewhat artificial-looking resolution problem that allows true induction on the dimension.

Problem 3.13.3. The end result of the resolution process guarantees that the birational transform of S has multiplicity < 2 along the birational transform of $H = (x = 0)$, but we have said nothing about the singularities that occur outside the birational transform of H.

There are indeed such singularities if we do not choose H carefully. For instance, if we take $H' := (x - z^2 = 0)$, then at the end of Step.1 of (3.11), that is, after two blow-ups, the birational transform of H' is $(x_2 - 1 = 0)$, which does not contain the singularity that is at the origin $(x_2 = y_2 = z_2 = 0)$.

Thus a careful choice of H is needed. This is solved by the theory of *maximal contact*, developed by Hironaka and Giraud [Gir74, AHV75].

Problem 3.13.4. In some cases, the opposite problem happens. All the singularities end up on the birational transforms of H, but we also pick up extra tangencies, so we see too many singularities.

For instance, take $H'' := (x - z^3 = 0)$. Since

$$x^2 + y^3 - z^6 = (x - z^3)(x + z^3) + y^3,$$

the trace of S on H'' is a triple line. The trace shows a 1-dimensional singular set when we have only an isolated singular point.

In other cases, these problems may appear only after many blow-ups.

At first glance, this may not be a problem at all. This simply means that we make some unnecessary blow-ups as well. Indeed, if our aim is to resolve surfaces only, then this problem can be mostly ignored. However, for the general inductive procedure this is a serious difficulty since unnecessary blow-ups can increase the multiplicity. For instance,

$$S = (x^4 + y^2 + yz^2 = 0) \subset \mathbb{A}^3$$

is an isolated double point. If we blow up the line $(x = y = 0)$, in the x-chart we get a triple point

$$x_1^3 + x_1 y_1^2 + y_1 z^2 = 0, \quad \text{where } x = x_1, y = y_1 x_1.$$

One way to solve this problem is to switch from resolving varieties to "resolving" ideal sheaves by introducing a *coefficient ideal* $C(S)$ such that

   (i) resolving S is equivalent to "resolving" $C(S)$, and
   (ii) "resolving" the traces $C(S)|_H$ does not generate extra blow-ups for S.

This change of emphasis is crucial for our approach.

Problem 3.13.5. No matter how carefully we choose H, we can never end up with a unique choice. For instance, the analytic automorphism of $S = (x^2 + y^3 - z^6 = 0)$,

$$(x, y, z) \mapsto (x + y^3, y\sqrt[3]{1 - 2x - y^3}, z),$$

shows that no internal property distinguishes the choice $x = 0$ from the choice $x + y^3 = 0$.

Even with the careful "maximal contact" choice of H, we end up with cases where the traces $S \cap H$ are not isomorphic. Thus our resolution process seems to depend on the choice of H.

This is again only a minor inconvenience for surfaces, but in higher dimensions we have to deal with patching together the local resolution processes into a global one. (We cannot even avoid this issue by pretending to care only about isolated singularities, since blowing up frequently leads to nonisolated singularities.)

An efficient solution of this problem developed in [Wło05] replaces S with an ideal $W(S)$ such that

(i) resolving S is equivalent to resolving $W(S)$,
(ii) the traces $W(S)|_H$ are locally isomorphic for all hypersurfaces of maximal contact through $s \in S$ (here "locally" is meant in the analytic or étale topology), and
(iii) the resolution of $W(S)|_H$ tells us how to resolve $W(S)$.

The local ambiguity is thus removed from the process, and there is no longer a patching problem.

Problem 3.13.6. At Steps 2 and 3 in (3.11), the choices we make are not canonical. For instance, in Step 2 we could have blown up the central curve with multiplicity 2 first, to complete the resolution in just one step. Even if we do Step 2 as above, in general there are many curves to blow-up in Step 3, and the order of blow-ups matters. (In $\mathbb{A}^3$, one can blow up two intersecting smooth curves in either order, and the resulting 3-folds are not isomorphic.)

This problem, too, remains unsolved. We make a choice, and it is good enough that the resolutions we get commute with any smooth morphism. Thus we get a resolution that one can call functorial. I would not call it a canonical resolution, since even in the framework of this proof other equally functorial choices are possible.

This is very much connected with the lack of minimal resolutions.

Next we see how Problems (3.13.2–5) can be approached for hypersurfaces using Weierstrass polynomials. As was the case with curves and surfaces, this example motivates the whole proof. (To be fair, this example provides much better guidance with hindsight. One might argue that the whole history of resolution by smooth blow-ups is but an ever-improving understanding of this single example. It has taken a long time to sort out how to generalize various aspects of it, and it is by no means certain that we have learned all the right lessons.)

EXAMPLE 3.14. Let $X \subset \mathbb{C}^{n+1}$ be a hypersurface. Pick a point $0 \in X$, where $\mathrm{mult}_0 X = m$. Choose suitable local coordinates $x_1, \ldots, x_n, z$, and apply the Weierstrass preparation theorem (1.93) to get (in an analytic neighborhood) an equation of the form

$$z^m + a_1(\mathbf{x})z^{m-1} + \cdots + a_m(\mathbf{x}) = 0$$

for X. We can kill the z^{m-1} term by a substitution $z = y - \frac{1}{m}a_1(\mathbf{x})$ to get another local equation

$$f := y^m + b_2(\mathbf{x})y^{m-2} + \cdots + b_m(\mathbf{x}) = 0. \tag{3.14.1}$$

Here $\mathrm{mult}_0 b_i \geq i$ since $\mathrm{mult}_0 X = m$.

Let us blow up the point 0 to get $\pi : B_0 X \to X$, and consider the chart $x_i' = x_i/x_n, x_n' = x_n, y' = y/x_n$. We get an equation for $B_0 X$

$$F := (y')^m + (x_n')^{-2}b_2(x_1'x_n', \ldots, x_n')(y')^{m-2} + \cdots + (x_n')^{-m}b_m(x_1'x_n', \ldots, x_n'). \tag{3.14.2}$$

Where are the points of multiplicity $\geq m$ on $B_0 X$? Locally we can view $B_0 X$ as a hypersurface in $\mathbb{C}^{n+1}$ given by the equation $F(\mathbf{x}', y') = 0$, and a point p has multiplicity $\geq m$ iff all the $(m-1)$st partials of F vanish. First of all, we get that

$$\frac{\partial^{m-1}F}{\partial y'^{m-1}} = m! \cdot y' \quad \text{vanishes at } p. \tag{3.14.3}$$

This means that all points of multiplicity $\geq m$ on $B_0 X$ are on the birational transform of the hyperplane $(y = 0)$. Since the new equation (3.14.2) has the same form as the original (3.14.1), the conclusion continues to hold after further blow-ups, solving (3.13.3):

Claim 3.14.4. After a sequence of blow-ups at points of multiplicity $\geq m$

$$\Pi : X_r = B_{p_{r-1}}X_{r-1} \to X_{r-1} = B_{p_{r-2}}X_{r-2} \to \cdots \to X_1 = B_{p_0}X \to X,$$

all points of multiplicity $\geq m$ on X_r are on the birational transform of the hyperplane $H := (y = 0)$, and all points of X_r have multiplicity $\leq m$.

This property of the hyperplane $(y = 0)$ will be encapsulated by the concept of *hypersurface of maximal contact*.

In order to determine the location of points of multiplicity m, we need to look at all the other $(m-1)$st partials of F restricted to $(y' = 0)$. These can be written as

$$\frac{\partial^{m-1}F}{\partial \mathbf{x}'^{i-1}\partial y'^{m-i}}\Big|_{(y'=0)} = (m-i)! \cdot \frac{\partial^{i-1}\big((x_n')^{-i}b_i(x_1'x_n', \ldots, x_n')\big)}{\partial \mathbf{x}'^{i-1}}. \tag{3.14.5}$$

Thus we can actually read off from $H = (y = 0)$ which points of $B_0 X$ have multiplicity m. For this, however, we need not only the restriction $f|_H = b_m(\mathbf{x})$ but all the other coefficients $b_i(\mathbf{x})$ as well.

There is one further twist. The usual rule for transforming a polynomial under a blow-up is

$$b(x_1, \ldots, x_n) \mapsto (x_n')^{-\operatorname{mult_0} b} b(x_1' x_n', \ldots, x_n'),$$

but instead we use the rule

$$b_i(x_1, \ldots, x_n) \mapsto (x_n')^{-i} b_i(x_1' x_n', \ldots, x_n').$$

That is, we "pretend" that b_i has multiplicity i at the origin. To handle this, we introduce the notion of a marked function (g, m) and define the birational transform of a marked function (g, m) to be

$$\pi_*^{-1}\big(g(x_1, \ldots, x_n), m\big) := \big((x_n')^{-m} g(x_1' x_n', \ldots, x_n'), m\big). \qquad (3.14.6)$$

Warning. If we change coordinates, the right-hand side of (3.14.6) changes by a unit. Thus the ideal $(\pi_*^{-1}(g, m))$ is well defined but not $\pi_*^{-1}(g, m)$ itself. Fortunately, this does not lead to any problems.

By induction we define $\Pi_*^{-1}(g, m)$, where Π is a sequence of blow-ups as in (3.14.4).

This leads to a solution of Problems (3.13.2) and (3.13.4).

Claim 3.14.7. After a sequence of blow-ups at points of multiplicity $\geq m$,

$$\Pi : X_r = B_{p_{r-1}} X_{r-1} \to X_{r-1} = B_{p_{r-2}} X_{r-2} \to \cdots \to X_1 = B_{p_0} X \to X,$$

a point $p \in X_r$ has multiplicity $< m$ on X_r iff

(i) either $p \notin H_r$, the birational transform of H,
(ii) or there is an index $i = i(p)$ such that

$$\operatorname{mult}_p (\Pi|_{H_r})_*^{-1} \big(b_i(\mathbf{x}), i\big) < i.$$

A further observation is that we can obtain the $b_i(\mathbf{x})$ from the derivatives of f:

$$b_i(\mathbf{x}) = \frac{1}{(m-i)!} \cdot \frac{\partial^{m-i} f}{\partial y^{m-i}}(\mathbf{x}, y)|_H.$$

Thus (3.14.7) can be restated in a more invariant-looking but also vaguer form.

Principle 3.14.8. Multiplicity reduction for the $n + 1$-variable function $f(\mathbf{x}, y)$ is equivalent to multiplicity reduction for certain n-variable functions constructed from the partial derivatives of f with suitable markings.

3.14.9. Until now we have completely ignored that everything we do depends on the initial choice of the coordinate system $(x_1, \ldots, x_n, z)$. The fact that in (3.14.7–8) we get equivalences suggests that the choice of the coordinate system should not matter much. The problem, however, remains: in globalizing the local resolutions constructed above, we have to choose local resolutions out of the many possibilities and hope that the different local choices patch together.

This has been a surprisingly serious obstacle.

3.3. Statement of the main theorems

So far we have been concentrating on resolution of singularities, but now we switch our focus, and instead of dealing with singular varieties, we consider ideal sheaves on *smooth* varieties. Given an ideal sheaf I on a smooth variety X, our first aim is to write down a birational morphism $g : X' \to X$ such that X' is smooth and the pulled-back ideal sheaf g^*I is locally principal. This is called the principalization of I.

NOTATION 3.15. Let $g : Y \to X$ be a morphism of schemes and $I \subset \mathcal{O}_X$ an ideal sheaf. I will be sloppy and use g^*I to denote the *inverse image ideal sheaf* of I. This is the ideal sheaf generated by the pull-backs of local sections of I. (It is denoted by $g^{-1}I \cdot \mathcal{O}_Y$ or by $I \cdot \mathcal{O}_Y$ in [Har77, Sec.II.7].)

We should be mindful that g^*I (as an inverse image ideal sheaf) may differ from the usual sheaf-theoretic pull-back, also commonly denoted by g^*I; see [Har77, II.7.12.2]. This can happen even if X, Y are both smooth.

For the rest of the chapter, we use only inverse image ideal sheaves, so hopefully this should not lead to any confusion.

It is easy to see that resolution of singularities implies principalization. Indeed, let $X_1 := B_I X$ be the blow-up of I with projection $\pi : X_1 \to X$. Then π^*I is locally principal (cf. [Har77, II.7.13]). Thus if $h : X' \to X_1$ is any resolution, then $\pi \circ h : X' \to X$ is a principalization of I.

Our aim, however, is to derive resolution theorems from principalization results. Given a singular variety Z, choose an embedding of Z into a smooth variety X, and let $I_Z \subset \mathcal{O}_X$ be its ideal sheaf. (For Z quasi-projective, we can just take any embedding $Z \hookrightarrow \mathbb{P}^N$ into a projective space, but in general such an embedding may not exist; see (3.46).) Then we turn a principalization of the ideal sheaf I_Z into a resolution of Z.

In this section we state four, increasingly stronger versions of principalization and derive from them various resolution theorems. The rest of the chapter is then devoted to proving these principalization theorems.

3.16 (Note on terminology). Principal ideals are much simpler than arbitrary ideals, but they can still be rather complicated since they capture all the intricacies of hypersurface singularities.

An ideal sheaf I on a smooth scheme X is called (locally) *monomial* if the following equivalent conditions hold.

(1) For every $x \in X$ there are local coordinates z_i and natural numbers c_i such that $I \cdot \mathcal{O}_{x,X} = \prod_i z_i^{c_i} \cdot \mathcal{O}_{x,X}$.
(2) I is the ideal sheaf of a simple normal crossing divisor (3.24).

I would like to call a birational morphism $g : X' \to X$, such that X' is smooth and g^*I is monomial, a resolution of I.

However, for many people, the phrase "resolution of an ideal sheaf" brings to mind a long exact sequence

$$\cdots \to E_2 \to E_1 \to I \to 0,$$

where the E_i are locally free sheaves. This has nothing to do with resolution of singularities. Thus, rather reluctantly, I follow convention and talk about *principalization* or *monomialization* of an ideal sheaf I.

We start with the simplest version of principalization (3.17) and its first consequence, the resolution of indeterminacies of rational maps (3.18). Then we consider a stronger version of principalization (3.21), which implies resolution of singularities (3.22). Monomialization of ideal sheaves is given in (3.26), which implies strong, functorial resolution for quasi-projective varieties (3.27). The proof of the strongest variant of monomialization (3.35) occupies the rest of the chapter. At the end of the section we observe that the functorial properties proved in (3.35) imply that the monomialization and resolution theorems automatically extend to algebraic and analytic spaces; see (3.42) and (3.44).

THEOREM 3.17 (Principalization, I). *Let X be a smooth variety over a field of characteristic zero and $I \subset \mathcal{O}_X$ a nonzero ideal sheaf. Then there is a smooth variety X' and a birational and projective morphism $f : X' \to X$ such that $f^*I \subset \mathcal{O}_{X'}$ is a locally principal ideal sheaf.*

COROLLARY 3.18 (Elimination of indeterminacies). *Let X be a smooth variety over a field of characteristic zero and $g : X \dashrightarrow \mathbb{P}$ a rational map to some projective space. Then there is a smooth variety X' and a birational and projective morphism $f : X' \to X$ such that the composite $g \circ f : X' \to \mathbb{P}$ is a morphism.*

Proof. Since $\mathbb{P}$ is projective and X is normal, there is a subset $Z \subset X$ of codimension ≥ 2 such that $g : X \setminus Z \to \mathbb{P}$ is a morphism. Thus $g^*\mathcal{O}_{\mathbb{P}}(1)$ is a line bundle on $X \setminus Z$. Since X is smooth, it extends to a line bundle on X; denote it by L. Let $J \subset L$ be the subsheaf generated by $g^*H^0(\mathbb{P}, \mathcal{O}_{\mathbb{P}}(1))$. Then $I := J \otimes L^{-1}$ is an ideal sheaf, and so by (3.17) there is a projective morphism $f : X' \to X$ such that $f^*I \subset \mathcal{O}_{X'}$ is a locally principal ideal sheaf.

Thus the global sections

$$(g \circ f)^*H^0(\mathbb{P}, \mathcal{O}_{\mathbb{P}}(1)) \subset H^0(X', f^*L)$$

generate the locally free sheaf $L' := f^*I \otimes f^*L$. Therefore, $g \circ f : X' \to \mathbb{P}$ is a morphism given by the nowhere-vanishing subspace of global sections

$$(g \circ f)^*H^0(\mathbb{P}, \mathcal{O}_{\mathbb{P}}(1)) \subset H^0(X', L'). \quad \square$$

NOTATION 3.19 (Blow-ups). Let X be a scheme and $Z \subset X$ a closed subscheme. Let $\pi = \pi_{Z,X} : B_Z X \to X$ denote the *blow-up* of Z in X; see [Sha94, II.4] or [Har77, Sec.II.7]. Although resolution by definition involves singular schemes X, we will almost always study the case where X and Z are both smooth, called a *smooth blow-up*. The *exceptional divisor* of a blow-up is $F := \pi_{Z,X}^{-1}(Z) \subset B_Z X$. If $\pi_{Z,X}$ is a smooth blow-up, then F and $B_Z X$ are both smooth.

WARNING 3.20 (Trivial and empty blow-ups). A blow-up is called *trivial* if Z is a Cartier divisor in X. In these cases $\pi_{Z,X} : B_Z X \to X$ is an isomorphism. We also allow the possibility $Z = \emptyset$, called the *empty* blow-up.

We have to deal with trivial blow-ups to make induction work since the blow-up of a codimension 2 smooth subvariety $Z^{n-2} \subset X^n$ corresponds to a trivial blow-up on a smooth hypersurface $Z^{n-2} \subset H^{n-1} \subset X^n$.

Two peculiarities of trivial blow-ups cause trouble.

(1) For a nontrivial smooth blow-up $\pi : B_Z X \to X$, the morphism π determines the center Z, but this fails for a trivial blow-up. One usually thinks of π as *the* blow-up, hiding the dependence on Z. By contrast, we always think of a smooth blow-up as having a specified center.

(2) The exceptional divisor of a trivial blow-up $\pi_{Z,X} : B_Z X \to X$ is $F = Z \subset X$. This is, unfortunately, at variance with the usual definition of exceptional set/divisor (see [Sha94, Sec.II.4.4] or (3.25)), but it is the right concept for blow-ups.

These are both minor inconveniences, but they could lead to confusion.

Empty blow-ups naturally occur when we restrict a blow-up sequence to an open subset $U \subset X$ and the center of the blow-up is disjoint from U. We will exclude empty blow-ups from the final blow-up sequences, but we have to keep them in mind since they mess up the numbering of the blow-up sequences.

THEOREM 3.21 (Principalization, II). *Let X be a smooth variety over a field of characteristic zero and $I \subset \mathcal{O}_X$ a nonzero ideal sheaf. Then there is a smooth variety X' and a birational and projective morphism $f : X' \to X$ such that*

(1) $f^*I \subset \mathcal{O}_{X'}$ *is a locally principal ideal sheaf,*

(2) $f : X' \to X$ *is an isomorphism over $X \setminus \text{cosupp} \, I$, where $\text{cosupp} \, I$ (or $\text{Supp}(\mathcal{O}_X/I)$) is the cosupport of I, and*

(3) f *is a composite of smooth blow-ups*

$$f : X' = X_r \xrightarrow{\pi_{r-1}} X_{r-1} \xrightarrow{\pi_{r-2}} \cdots \xrightarrow{\pi_1} X_1 \xrightarrow{\pi_0} X_0 = X.$$

This form of principalization implies resolution of singularities, seemingly by accident. (In practice, one can follow the steps of a principalization method and see how resolution happens, though this is not always easy.)

COROLLARY 3.22 (Resolution of singularities, I). *Let X be a quasi-projective variety. Then there is a smooth variety X' and a birational and projective morphism $g : X' \to X$.*

Proof. Choose an embedding of X into a smooth variety P such that $N \geq \dim X + 2$. (For instance, $P = \mathbb{P}^N$ works for all $N \gg \dim X$.) Let $\bar{X} \subset P$ denote the closure and $I \subset \mathcal{O}_P$ its ideal sheaf. Let $\eta_X \in X \subset P$ be the generic point of X.

By (3.21), there is a sequence of smooth blow-ups

$$\Pi : P' = P_r \xrightarrow{\pi_{r-1}} P_{r-1} \xrightarrow{\pi_{r-2}} \cdots P_1 \xrightarrow{\pi_0} P_0 = P$$

such that Π^*I is locally principal.

Since X has codimension ≥ 2, its ideal sheaf I is not locally principal at η_X, and therefore, some blow-up center must contain η_X. Thus, there is a unique j such that $\pi_0 \cdots \pi_{j-1} : P_j \to P$ is a local isomorphism around η_X but $\pi_j : P_{j+1} \to P_j$ is a blow-up with center $Z_j \subset P_j$ such that $\eta_X \in Z_j$.

By (3.21.2), $\pi_0 \cdots \pi_{j-1}(Z_j) \subset \bar{X}$, and this implies that η_X is the generic point of Z_j. Thus

$$g := \pi_0 \cdots \pi_{j-1} : Z_j \to \bar{X}$$

is birational.

Z_j is smooth since we blow it up, and by (3.21.3) we only blow up smooth subvarieties. Therefore g is a resolution of singularities of $\bar{X}$. Set $X' := g^{-1}(X) \subset Z_j$. Then $g : X' \to X$ is a resolution of singularities of X. □

WARNING 3.23. The resolution $g : X' \to X$ constructed in (3.22) need not be a composite of *smooth* blow-ups. Indeed, the process exhibits g as the composite of blow-ups whose centers are obtained by intersecting the smooth centers Z_i with the birational transforms of X. Such intersections may be singular. See (3.106) for a concrete example.

We also need a form of resolution that keeps track of a suitable simple normal crossing divisor. This feature is very useful in applications and in the inductive proof.

DEFINITION 3.24. Let X be a smooth variety and $E = \sum E^i$ a *simple normal crossing divisor* on X. This means that each E^i is smooth, and for each point $x \in X$ one can choose local coordinates $z_1, \ldots, z_n \in m_x$ in the maximal ideal of the local ring $\mathcal{O}_{x,X}$ such that for each i

(1) either $x \notin E^i$, or
(2) $E^i = (z_{c(i)} = 0)$ in a neighborhood of x for some $c(i)$, and

(3) $c(i) \neq c(i')$ if $i \neq i'$.

A subvariety $Z \subset X$ has *simple normal crossings* with E if one can choose $z_1, \ldots, z_n$ as above such that in addition

(4) $Z = (z_{j_1} = \cdots = z_{j_s} = 0)$ for some $j_1, \ldots, j_s$, again in some open neighborhood of x.

In particular, Z is smooth, and some of the E^i are allowed to contain Z.

If E does not contain Z, then $E|_Z$ is again a simple normal crossing divisor on Z.

DEFINITION 3.25. Let $g : X' \to X$ be a birational morphism. Its *exceptional set* is the set of points $x' \in X'$ such that g is not a local isomorphism at x'. It is denoted by $\mathrm{Ex}(g)$. If X is smooth, then $\mathrm{Ex}(g)$ is a divisor [Sha94, II.4.4]. Let

$$\Pi : X' = X_r \xrightarrow{\pi_{r-1}} X_{r-1} \xrightarrow{\pi_{r-2}} \cdots \xrightarrow{\pi_1} X_1 \xrightarrow{\pi_0} X_0 = X$$

be a sequence of smooth blow-ups with centers $Z_i \subset X_i$. Define the *total exceptional set* to be

$$\mathrm{Ex}_{\mathrm{tot}}(\Pi) := \bigcup_{i=0}^{r-1} (\pi_i \circ \cdots \circ \pi_{r-1})^{-1}(Z_i).$$

If all the blow-ups are nontrivial, then $\mathrm{Ex}(\Pi) = \mathrm{Ex}_{\mathrm{tot}}(\Pi)$.

Let E be a simple normal crossing divisor on X. We say that the centers Z_i have *simple normal crossings with E* if each blow-up center $Z_i \subset X_i$ has simple normal crossings (3.24) with

$$(\pi_0 \cdots \pi_{i-1})_*^{-1}(E) + \mathrm{Ex}_{\mathrm{tot}}(\pi_0 \cdots \pi_{i-1}).$$

If this holds, then

$$\Pi_{\mathrm{tot}}^{-1}(E) := \Pi_*^{-1}(E) + \mathrm{Ex}_{\mathrm{tot}}(\Pi)$$

is a simple normal crossing divisor, called the *total transform* of E. (A refinement for divisors with ordered index set will be introduced in (3.65).)

We can now strengthen the theorem on principalization of ideal sheaves.

THEOREM 3.26 (Principalization, III). *Let X be a smooth variety over a field of characteristic zero, $I \subset \mathcal{O}_X$ a nonzero ideal sheaf and E a simple normal crossing divisor on X. Then there is a sequence of smooth blow-ups*

$$\Pi : R_{I,E}(X) := X_r \xrightarrow{\pi_{r-1}} X_{r-1} \xrightarrow{\pi_{r-2}} \cdots \xrightarrow{\pi_1} X_1 \xrightarrow{\pi_0} X_0 = X$$

whose centers have simple normal crossing with E such that

(1) $\Pi^* I \subset \mathcal{O}_{R_{I,E}(X)}$ *is the ideal sheaf of a simple normal crossing divisor, and*

(2) $\Pi : R_{I,E}(X) \to X$ *is functorial on smooth morphisms (3.4).*

Note that since Π is a composite of smooth blow-ups, $R_{I,E}(X)$ is smooth and $\Pi : R_{I,E}(X) \to X$ is birational and projective.

As a consequence we get strong resolution of singularities for quasi-projective schemes over a field of characteristic zero.

THEOREM 3.27 (Resolution of singularities, II). *Let X be a quasi-projective variety over a field of characteristic zero. Then there is a birational and projective morphism $\Pi : R(X) \to X$ such that*

(1) *$R(X)$ is smooth,*

(2) *$\Pi : R(X) \to X$ is an isomorphism over the smooth locus X^{ns}, and*

(3) *$\Pi^{-1}(\text{Sing } X)$ is a divisor with simple normal crossing.*

Proof. We have already seen in (3.22) that given a (locally closed) embedding $i : X \hookrightarrow P$ we get a resolution $R(X) \to X$ from the principalization of the ideal sheaf I of the closure of $i(X)$. We need to check that applying (3.26) to $(P, I, \emptyset)$ gives a strong resolution of X. (We do not claim that $R(X) \to X$ is independent of the embedding $i : X \hookrightarrow P$. This will have to wait until after the stronger principalization theorem (3.35).)

As in the proof of (3.22), there is a sequence of smooth blow-ups

$$\Pi : P' = P_r \xrightarrow{\pi_{r-1}} P_{r-1} \xrightarrow{\pi_{r-2}} \cdots P_1 \xrightarrow{\pi_0} P_0 = P$$

such that $\Pi^* I$ is locally principal. Moreover, there is a first blow-up

$$\pi_j : P_{j+1} \to P_j \quad \text{with center } Z_j \subset P_j$$

such that $g := \pi_0 \cdots \pi_{j-1}|_{Z_j} : Z_j \to \bar{X}$ is birational. We claim that $g : Z_j \to \bar{X}$ is a strong resolution of $\bar{X}$, and hence $g : g^{-1}(X) \to X$ is a strong resolution of X.

First, we prove that g is an isomorphism over $\bar{X}^{ns}$. As in (3.4.2), this follows from the functoriality condition (3.26.2). Note, however, that (3.26.2) asserts functoriality for $\Pi : P_r \to P$ but not for the intermediate maps $P_j \to P$. Thus a little extra work is needed. (This seems like a small technical point, but actually it has been the source of serious troubles. The notion of blow-up sequence functors (3.31) is designed to deal with it.)

Let $F'_{j+1} \subset P_r$ denote the birational transform of $F_{j+1} \subset P_{j+1}$, the exceptional divisor of π_j. Since $F'_{j+1} \subset \text{Ex}_{\text{tot}}(\Pi)$, it is a smooth divisor and so $\Pi|_{F'_{j+1}} : F'_{j+1} \to \bar{X}$ is generically smooth. Thus there is a smooth point $x \in X$ such that $\Pi|_{F'_{j+1}}$ is smooth over x.

For any other smooth point $x' \in X$, the embeddings

$$(x \in X \hookrightarrow P) \quad \text{and} \quad (x' \in X \hookrightarrow P)$$

have isomorphic étale neighborhoods. Thus by (3.26.2), $\Pi|_{F'_{j+1}}$ is also smooth over x'. We can factor

$$\Pi|_{F'_{j+1}} : F'_{j+1} \to Z_j \xrightarrow{g} \bar{X}.$$

Thus $g : Z_j \to \bar{X}$ is smooth over every smooth point of $\bar{X}$. It is also birational, and thus g is an isomorphism over $\bar{X}^{ns}$.

Since Z_j is smooth, g is not an isomorphism over any point of Sing $\bar{X}$, and thus

$$g^{-1}(\text{Sing }\bar{X}) = Z_j \cap \text{Ex}_{\text{tot}}(\pi_0 \cdots \pi_{j-1}),$$

where Ex_{tot} denotes the total exceptional divisor (3.25). Observe that in (3.26) we can only blow-up Z_j if it has simple normal crossings with $\text{Ex}_{\text{tot}}(\pi_0 \cdots \pi_{j-1})$; hence

$$g^{-1}(\text{Sing }\bar{X}) = Z_j \cap \text{Ex}_{\text{tot}}(\pi_0 \cdots \pi_{j-1})$$

is a simple normal crossing divisor on Z_j. □

REMARK 3.28. The proof of the implication (3.26) ⇒ (3.27) also works for any scheme that can be embedded into a smooth variety. We see in (3.46) that not all schemes can be embedded into a smooth scheme, so in general one has to proceed differently. It is worthwhile to contemplate further the local nature of resolutions and its consequences.

Let X be a scheme of finite type and $X = \cup U_i$ an affine cover. For each U_i (3.27) gives a resolution $R(U_i) \to U_i$, and we would like to patch these together to $R(X) \to X$.

First, we need to show that $R(U_i)$ is well defined; that is, it does not depend on the embedding $i : U_i \hookrightarrow P$ chosen in the proof of (3.27).

Second, we need to show that $R(U_i)$ and $R(U_j)$ agree over the intersection $U_i \cap U_j$.

If these hold, then the $R(U_i)$ patch together into a resolution $R(X) \to X$, but there is one problem. $R(X) \to X$ is *locally* projective, but it may not be *globally* projective. The following is an example of this type.

Example 3.28.1. Let X be a smooth 3-fold and C_1, C_2 a pair of irreducible curves, intersecting at two points p_1, p_2. Assume, furthermore, that C_i is smooth away from p_i, where it has a cusp whose tangent plane is transversal to the other curve. Let $I \subset \mathcal{O}_X$ be the ideal sheaf of $C_1 \cup C_2$.

On $U_1 = X \setminus \{p_1\}$, the curve C_1 is smooth; we can blow it up first. The birational transform of C_2 becomes smooth, and we can blow it up next to get $Y_1 \to U_1$. Over $U_2 = X \setminus \{p_2\}$ we would work in the other order. Over $U_1 \cap U_2$ we get the same thing, and thus Y_1 and Y_2 glue together to a variety Y such that $Y \to X$ is proper and locally projective but not globally projective.

We see that the gluing problem comes from the circumstance that the birational map $Y_1 \cap Y_2 \to U_1 \cap U_2$ is the blow-up of two disjoint curves, and we do not know which one to blow up first.

For a sensible resolution algorithm there is only one choice: we have to blow them up at the same time. Thus in the above example, the "correct" method is to blow up the points p_1, p_2 first. The curves C_1, C_2 become smooth and disjoint, and then both can be blown up. (More blow-ups are needed if we want to have only simple normal crossings.)

These problems can be avoided if we make (3.26) sharper. A key point is to prove functoriality conditions not only for the end result $R_{I,E}(X)$ but for all intermediate steps, including the center of each blow-up.

DEFINITION 3.29 (Blow-up sequences). Let X be a scheme. A *blow-up sequence* of length r starting with X is a chain of morphisms

$$\Pi : X_r \xrightarrow{\pi_{r-1}} X_{r-1} \xrightarrow{\pi_{r-2}} \cdots \xrightarrow{\pi_1} X_1 \xrightarrow{\pi_0} X_0 = X, \tag{3.29.1}$$

$$\cup \qquad\qquad \cup \qquad \cup$$

$$Z_{r-1} \qquad \cdots \qquad Z_1 \qquad Z_0$$

where each $\pi_i = \pi_{Z_i,X_i} : X_{i+1} \to X_i$ is a blow-up with center $Z_i \subset X_i$ and exceptional divisor $F_{i+1} \subset X_{i+1}$. Set

$$\Pi_{ij} := \pi_j \circ \cdots \circ \pi_{i-1} : X_i \to X_j \quad \text{and} \quad \Pi_i := \Pi_{i0} : X_i \to X_0.$$

We say that (3.29.1) is a *smooth blow-up sequence* if each $\pi_i : X_{i+1} \to X_i$ is a smooth blow-up.

We allow trivial and empty blow ups (3.20).

For the rest of the chapter, π always denotes a blow-up, Π_{ij} a composite of blow-ups and Π the composite of all blow-ups in a blow-up sequence (whose length we frequently leave unspecified). We usually drop the centers Z_i from the notation, to avoid cluttering up the diagrams.

DEFINITION 3.30 (Transforming blow-up sequences). There are three basic ways to transform blow-up sequences from one scheme to another. Let $\mathbf{B} :=$

$$\Pi : X_r \xrightarrow{\pi_{r-1}} X_{r-1} \xrightarrow{\pi_{r-2}} \cdots \xrightarrow{\pi_1} X_1 \xrightarrow{\pi_0} X_0 = X$$

$$\cup \qquad\qquad \cup \qquad \cup$$

$$Z_{r-1} \qquad \cdots \qquad Z_1 \qquad Z_0$$

be a blow-up sequence starting with X.

3.30.1. For a smooth morphism $h : Y \to X$ define the *pull-back* $h^*\mathbf{B}$ to be the blow-up sequence

$$h^*\Pi : X_r \times_X Y \xrightarrow{h^*\pi_{r-1}} X_{r-1} \times_X Y \cdots X_1 \times_X Y \xrightarrow{h^*\pi_0} X_0 \times_X Y = Y.$$

$$\cup \qquad\qquad \cup \qquad \cup$$

$$Z_{r-1} \times_X Y \cdots Z_1 \times_X Y \qquad Z_0 \times_X Y$$

If $\mathbf{B}$ is a smooth blow-up sequence then so is $h^*\mathbf{B}$. If h is surjective then $h^*\mathbf{B}$ determines $\mathbf{B}$ uniquely. However, if h is not surjective, then $h^*\mathbf{B}$ may contain some empty blow-ups, and we lose information about the centers living above $X \setminus h(Y)$.

3.30.2. Let X be a scheme and $j : S \hookrightarrow X$ a closed subscheme. Given a blow-up sequence $\mathbf{B}$ starting with X as above, define its *restriction* to S as the sequence

$$
\Pi^S : S_r \xrightarrow{\pi^S_{r-1}} S_{r-1} \xrightarrow{\pi^S_{r-2}} \cdots \xrightarrow{\pi^S_1} S_1 \xrightarrow{\pi^S_0} S_0 = S.
$$
$$
\cup \qquad\qquad\qquad\qquad \cup \qquad\qquad \cup
$$
$$
Z_{r-1} \cap S_{r-1} \qquad \cdots \qquad Z_1 \cap S_1 \qquad Z_0 \cap S_0
$$

It is denoted by $j^*\mathbf{B}$ or $\mathbf{B}|_S$.

Note that $S_{i+1} := B_{Z_i \cap S_i} S_i$ is naturally identified with the birational transform $(\pi_i)^{-1}_* S_i \subset X_{i+1}$ (cf. [Har77, II.7.15]), thus there are natural embeddings $S_i \hookrightarrow X_i$ for every i.

The restriction of a smooth blow-up sequence need not be a smooth blow-up sequence.

3.30.3. Conversely, let $\mathbf{B}(S) :=$

$$
\Pi := S_r \xrightarrow{\pi_{r-1}} S_{r-1} \xrightarrow{\pi_{r-2}} \cdots \xrightarrow{\pi_1} S_1 \xrightarrow{\pi_0} S_0 = S
$$

be a blow-up sequence with centers $Z^S_i \subset S_i$. Define its *push-forward* as the sequence $j_*\mathbf{B} :=$

$$
\Pi^X : X_r \xrightarrow{\pi^X_{r-1}} X_{r-1} \xrightarrow{\pi^X_{r-2}} \cdots \xrightarrow{\pi^X_1} X_1 \xrightarrow{\pi^X_0} X_0 = X,
$$

whose centers $Z^X_i \subset X_i$ are defined inductively as $Z^X_i := (j_i)_* Z^S_i$, where the $j_i : S_i \hookrightarrow X_i$ are the natural inclusions. Thus, for all practical purposes, $Z^X_i = Z^S_i$.

If $\mathbf{B}$ is a smooth blow-up sequence, then so is $j_*\mathbf{B}$.

DEFINITION 3.31 (Blow-up sequence functors). A *blow-up sequence functor* is a functor $\mathcal{B}$ whose

(1) inputs are triples (X, I, E), where X is a scheme, $I \subset \mathcal{O}_X$ an ideal sheaf that is nonzero on every irreducible component and E a divisor on X with ordered index set, and

(2) outputs are blow-up sequences

$$
\Pi : X_r \xrightarrow{\pi_{r-1}} X_{r-1} \xrightarrow{\pi_{r-2}} \cdots \xrightarrow{\pi_1} X_1 \xrightarrow{\pi_0} X_0 = X
$$
$$
\cup \qquad\qquad\qquad\qquad \cup \qquad\qquad \cup
$$
$$
Z_{r-1} \qquad \cdots \qquad Z_1 \qquad Z_0
$$

with specified centers. Here the length of the sequence r, the schemes X_i and the centers Z_i all depend on (X, I, E). (Later we will add ideal sheaves I_i and divisors E_i to the notation.)

If each Z_i is smooth, then a nontrivial blow-up $\pi_i : X_{i+1} \to X_i$ uniquely determines Z_i, so we can drop Z_i from the notation. However, in general many different centers give the same birational map.

The (partial) *resolution functor* $\mathcal{R}$ *associated* to a blow-up sequence functor $\mathcal{B}$ is the functor that sends (X, I, E) to the end result of the blow-up sequence

$$\mathcal{R} : (X, I, E) \mapsto (\Pi : X_r \to X).$$

Sometimes we write simply $\mathcal{R}_{(I,E)}(X) = X_r$.

3.32 (Empty blow-up convention). We basically try to avoid empty blow-ups, but we are forced to deal with them because a pull-back or a restriction of a nonempty blow-up may be an empty blow-up.

Instead of saying repeatedly that we perform a certain blow-up unless its center is empty, we adopt the convention that the final outputs of the named blow-up sequence functors $\mathcal{BD}, \mathcal{BMO}, \mathcal{BO}, \mathcal{BP}$ do not contain empty blow-ups.

The process of their construction may contain blow-ups that are empty in certain cases. (For instance, we may be told to blow up $E^1 \cap E^2$ and the intersection may be empty.) These steps are then ignored without explicit mention whenever they happen to lead to empty blow-ups.

REMARK 3.33. The end result of a sequence of blow-ups $\Pi : X_r \to X$ often determines the whole sequence, but this is not always the case.

First, there are some genuine counterexamples. Let $p \in C$ be a smooth pointed curve in a smooth 3-fold X_0. We can first blow up p and then the birational transform of C to get

$$\Pi : X_2 \xrightarrow{\pi_1} X_1 = B_p X_0 \xrightarrow{\pi_0} X_0,$$

with exceptional divisors $E_0, E_1 \subset X_2$, or we can blow up first C and then the preimage $D = \sigma_0^{-1}(p)$ to get

$$\Sigma : X_2' \xrightarrow{\sigma_1} X_1' = B_C X_0 \xrightarrow{\sigma_0} X_0$$

with exceptional divisors $E_0', E_1' \subset X_2'$.

It is easy to see that $X_2 \cong X_2'$, and under this isomorphism E_1 corresponds to E_0' and E_0 corresponds to E_1'.

Second, there are some "silly" counterexamples. If $Z_1, Z_2 \subset X$ are two disjoint smooth subvarieties, then we get the same result whether we blow up first Z_1 and then Z_2, or first Z_2 and then Z_1, or in one step we blow up $Z_1 \cup Z_2$.

While it seems downright stupid to distinguish between these three processes, it is precisely this ambiguity that caused the difficulties in (3.28.1).

It is also convenient to have a unified way to look at the functoriality properties of various resolutions.

3.34 (Functoriality package). There are three functoriality properties of blow-up sequence functors $\mathcal{B}$ that we are interested in. Note that in all

three cases the claimed isomorphism is unique, and hence the existence is
a local question.

Functoriality for étale morphisms is an essential ingredient of the proof.
As noted in (3.9.2), this is equivalent to functoriality for smooth morphisms
(3.34.1). Independence of the base field (3.34.2) is very useful in applications, but it is not needed for the proofs.

Functoriality for closed embeddings (3.34.3) is used for resolution of
singularities, but it is not needed for the principalization theorems. This
property is quite delicate, and we are not able to prove it in full generality,
see (3.71).

3.34.1 (Smooth morphisms). We would like our resolutions to commute with smooth morphisms, and it is best to build this into the blow-up
sequence functors.

We say that a blow-up sequence functor $\mathcal{B}$ commutes with h if

$$\mathcal{B}\big(Y, h^*I, h^{-1}(E)\big) = h^* \mathcal{B}(X, I, E).$$

This sounds quite reasonable until one notices that even when $Y \to X$ is
an open immersion it can happen that $Z_0 \times_X Y$ is empty. It is, however,
reasonable to expect that a good blow-up sequence functor commutes with
smooth *surjections*.

Therefore, we say that $\mathcal{B}$ commutes with smooth morphisms if

- $\mathcal{B}$ commutes with every smooth surjection h, and
- for every smooth morphism h, $\mathcal{B}(Y, h^*I, h^{-1}(E))$ is obtained from
  the pull-back $h^* \mathcal{B}(X, I, E)$ by deleting every blow-up $h^*\pi_i$ whose
  center is empty and reindexing the resulting blow-up sequence.

3.34.2 (Change of fields). We also would like the resolution to be independent of the field we work with.

Let $\sigma : K \hookrightarrow L$ be a field extension. Given a K-scheme of finite type
$X_K \to \operatorname{Spec} K$, we can view $\operatorname{Spec} L$ as a scheme over $\operatorname{Spec} K$ (possibly not
of finite type) and take the fiber product

$$X_{L,\sigma} := X_K \times_{\operatorname{Spec} K} \operatorname{Spec} L,$$

which is an L-scheme of finite type. If I is an ideal sheaf and E a divisor
on X, then similarly we get $I_{L,\sigma}$ and $E_{L,\sigma}$.

We say that $\mathcal{B}$ *commutes* with σ if $\mathcal{B}(X_{L,\sigma}, I_{L,\sigma}, E_{L,\sigma})$ is the blow-up
sequence

$$\Pi_{L,\sigma} : (X_r)_{L,\sigma} \xrightarrow{(\pi_{r-1})_{L,\sigma}} (X_{r-1})_{L,\sigma} \quad \cdots \quad (X_1)_{L,\sigma} \xrightarrow{(\pi_0)_{L,\sigma}} (X_0)_{L,\sigma}.$$
$$\cup \qquad\qquad\qquad \cup \qquad\qquad\quad \cup$$
$$(Z_{r-1})_{L,\sigma} \quad \cdots \quad (Z_1)_{L,\sigma} \qquad\qquad (Z_0)_{L,\sigma}$$

This property will hold automatically for all blow-up sequence functors that
we construct.

3.34.3 (Closed embeddings). In the proof of (3.22) we constructed a resolution of a variety Z by choosing an embedding of Z into a smooth variety Y. In order to get a well-defined resolution, we need to know that our constructions do not depend on the embedding chosen. The key step is to ensure independence from further embeddings $Z \hookrightarrow Y \hookrightarrow X$.

We say that $\mathcal{B}$ *commutes with closed embeddings* if

$$\mathcal{B}(X, I_X, E) = j_* \mathcal{B}(Y, I_Y, E|_Y),$$

whenever

- $j : Y \hookrightarrow X$ is a closed embedding of smooth schemes,
- $0 \neq I_Y \subset \mathcal{O}_Y$ and $0 \neq I_X \subset \mathcal{O}_X$ are ideal sheaves such that $\mathcal{O}_X/I_X = j_*(\mathcal{O}_Y/I_Y)$, and
- E is a simple normal crossing divisor on X such that $E|_Y$ is also a simple normal crossing divisor on Y.

3.34.4 (Closed embeddings, weak form). Let the notation and assumptions be as in (3.34.3). We say that $\mathcal{B}$ *weakly commutes with closed embeddings* if

$$j^* \mathcal{B}(X, I_X, E) = \mathcal{B}(Y, I_Y, E|_Y).$$

The difference appears only in the proof of (3.35) given in (3.72). At the beginning of the proof we blow up various intersections of the irreducible components of E. Since these intersections are not contained in Y, this commutes with restriction to Y but it does not commute with push forward.

The strongest form of monomialization is the following.

THEOREM 3.35 (Principalization, IV). *There is a blow-up sequence functor $\mathcal{BP}$ defined on all triples (X, I, E), where X is a smooth scheme of finite type over a field of characteristic zero, $I \subset \mathcal{O}_X$ is an ideal sheaf that is not zero on any irreducible component of X and E is a simple normal crossing divisor on X. $\mathcal{BP}$ satisfies the following conditions.*

(1) *In the blow-up sequence $\mathcal{BP}(X, I, E) =$*

$$\Pi : X_r \xrightarrow{\pi_{r-1}} X_{r-1} \xrightarrow{\pi_{r-2}} \cdots \xrightarrow{\pi_1} X_1 \xrightarrow{\pi_0} X_0 = X,$$
$$\cup \qquad\qquad\qquad \cup \qquad\quad \cup$$
$$Z_{r-1} \qquad \cdots \qquad Z_1 \qquad\quad Z_0$$

all centers of blow-ups are smooth and have simple normal crossing with E (3.25).

(2) *The pull-back $\Pi^* I \subset \mathcal{O}_{X_r}$ is the ideal sheaf of a simple normal crossing divisor.*

(3) *$\Pi : X_r \to X$ is an isomorphism over $X \setminus \mathrm{cosupp}\, I$.*

(4) *$\mathcal{BP}$ commutes with smooth morphisms (3.34.1) and with change of fields (3.34.2).*

(5) *$\mathcal{BP}$ commutes with closed embeddings (3.34.3) whenever $E = \emptyset$.*

Putting together the proof of (3.27) with (3.37), we obtain strong and functorial resolution.

THEOREM 3.36 (Resolution of singularities, III). *There is a blow-up sequence functor* $\mathcal{BR}(X) =$

$$\Pi : X_r \xrightarrow{\pi_{r-1}} X_{r-1} \xrightarrow{\pi_{r-2}} \cdots \xrightarrow{\pi_1} X_1 \xrightarrow{\pi_0} X_0 = X,$$

with the vertical inclusions $Z_{r-1}, \cdots, Z_1, Z_0$

defined on all schemes X of finite type over a field of characteristic zero, satisfying the following conditions.

(1) X_r *is smooth.*
(2) $\Pi : X_r \to X$ *is an isomorphism over the smooth locus* X^{ns}.
(3) $\Pi^{-1}(\operatorname{Sing} X)$ *is a divisor with simple normal crossings.*
(4) $\mathcal{BR}$ *commutes with smooth morphisms (3.34.1) and with change of fields (3.34.2).*

Proof. First we construct $\mathcal{BR}(X)$ for affine schemes. Pick any embedding $X \hookrightarrow A$ into a smooth affine scheme such that $\dim A \geq \dim X + 2$. As in the proof of (3.22), the blow-up sequence for $\mathcal{BP}(A, I_X, \emptyset)$ obtained in (3.35) gives a blow-up sequence $\mathcal{BR}(X)$.

Before we can even consider the functoriality conditions, we need to prove that $\mathcal{BR}(X)$ is independent of the choice of the embedding $X \hookrightarrow A$.

Thus assume that $\Pi_1 : R_1(X) \to \cdots \to X$ and $\Pi_2 : R_2(X) \to \cdots \to X$ are two blow-up sequences constructed this way. Using that $\mathcal{BP}$ weakly commutes with closed embeddings (3.34.4), it is enough to prove uniqueness for resolutions constructed from embeddings into affine spaces $X \hookrightarrow \mathbb{A}^n$. Moreover, we are allowed to increase n anytime by taking a further embedding $\mathbb{A}^n \hookrightarrow \mathbb{A}^{n+m}$.

As (3.39) shows, any two embeddings $i_1, i_2 : X \hookrightarrow \mathbb{A}^n$ become equivalent by an automorphism of $\mathbb{A}^{2n}$, which gives the required uniqueness.

Thus (3.34.2) for $\mathcal{BP}(A, I_X, \emptyset)$ implies (3.34.2) for $\mathcal{BR}(X)$ since an embedding $i : X \hookrightarrow A$ over K and $\sigma : K \hookrightarrow L$ gives another embedding $i_{\sigma,L} : X_{\sigma,L} \hookrightarrow A_{\sigma,L}$.

We can also reduce the condition (3.34.1) for $\mathcal{BR}(X)$ to the same condition for $\mathcal{BP}(A, I_X, \emptyset)$.

To see this, let $h : Y \to X$ be a smooth morphism, and choose any embedding $X \hookrightarrow A_X$ into a smooth affine variety. By (3.41), for every $y \in Y$ there is an open neighborhood $h(y) \in A_X^0 \subset A_X$ and a smooth surjection $h_A : A_Y^0 \twoheadrightarrow A_X^0$ such that $h_A^{-1}(X \cap A_X^0)$ is isomorphic to an open neighborhood $y \in Y^0 \subset Y$. Set $X^0 := X \cap A_X^0$. Thus, by (3.34.1),

$$h_A^* \mathcal{BP}(A_X^0, I_{X^0}, \emptyset) = \mathcal{BP}(A_Y^0, I_{Y^0}, \emptyset),$$

which shows that $h^* \mathcal{BR}(X^0) = \mathcal{BR}(Y^0)$. As we noted earlier, (3.34.1) is a local property, and thus $h^* \mathcal{BR}(X) = \mathcal{BR}(Y)$ as required.

We have now defined $\mathcal{BR}$ on (possibly reducible) affine schemes, and it remains to prove that one can glue together a global resolution out of these local pieces. This turns out to be a formal property of blow-up sequence functors, which we treat next.   □

PROPOSITION 3.37. *Let $\mathcal{B}$ be a blow-up sequence functor defined on affine schemes over a field k that commutes with smooth surjections.*

Then $\mathcal{B}$ has a unique extension to a blow-up sequence functor $\overline{\mathcal{B}}$, which is defined on all schemes of finite type over k and which commutes with smooth surjections.

Proof. For any X choose an open affine cover $X = \cup U_i$, and let $X' := \coprod_i U_i$ be the disjoint union. Then X' is affine, and there is a smooth surjection $g : X' \to X$. We show that $\mathcal{B}(X')$ descends to a blow-up sequence of X.

Set $X'' := \coprod_{i \leq j} U_i \cap U_j$. (We can also think of it as the fiber product $X' \times_X X'$.) There are surjective open immersions $\tau_1, \tau_2 : X'' \to X'$, where $\tau_1|_{U_i \cap U_j} : U_i \cap U_j \to U_i$ is the first inclusion and $\tau_2|_{U_i \cap U_j} : U_i \cap U_j \to U_j$ is the second.

The blow-up sequence $\mathcal{B}(X')$ starts with blowing up $Z'_0 \subset X'$, and the blow-up sequence $\mathcal{B}(X'')$ starts with blowing up $Z''_0 \subset X''$. Since $\mathcal{B}$ commutes with the τ_i, we conclude that

$$\tau_1^*(Z'_0) = Z''_0 = \tau_2^*(Z'_0). \tag{3.37.1}$$

Since $Z'_0 \subset X'$ is a disjoint union of its pieces $Z'_{0i} := Z'_0 \cap U_i$, (3.37.1) is equivalent to saying that for every i, j

$$Z'_{0i}|_{U_i \cap U_j} = Z'_{0j}|_{U_i \cap U_j}. \tag{3.37.2}$$

Thus the subschemes $Z'_{0i} \subset U_i$ glue together to a subscheme $Z_0 \subset X$.

This way we obtain $X_1 := B_{Z_0} X$ such that $X'_1 = X' \times_X X_1$. We can repeat the above argument to obtain the center $Z_1 \subset X_1$ and eventually get the whole blow-up sequence for X.   □

WARNING 3.38. A key element of the above argument is that we need to know $\mathcal{B}$ for the disconnected affine scheme $\coprod_i U_i$.

Any resolution functor defined on connected schemes automatically extends to disconnected schemes, but for blow-up sequence functors this is not at all the case. Although the blow-ups on different connected components do not affect each other, in a resolution process we need to know in which order we perform them, see (3.28.1).

Besides proving resolution for nonprojective schemes and for algebraic spaces, the method of (3.37) is used in the proof of the principalization theorems. The inductive proof naturally produces resolution processes only locally, and this method shows that they automatically globalize.

The following lemma shows that an affine scheme has a unique embedding into affine spaces, if we stabilize the dimension.

LEMMA 3.39. *Let X be an affine scheme and $i_1 : X \hookrightarrow \mathbb{A}^n$ and $i_2 : X \hookrightarrow \mathbb{A}^m$ two closed embeddings. Then the two embeddings into the coordinate subspaces*

$$i_1' : X \hookrightarrow \mathbb{A}^n \hookrightarrow \mathbb{A}^{n+m} \quad and \quad i_2' : X \hookrightarrow \mathbb{A}^m \hookrightarrow \mathbb{A}^{n+m}$$

are equivalent under a (nonlinear) automorphism of $\mathbb{A}^{n+m}$.

Proof. We can extend i_1 to a morphism $j_1 : \mathbb{A}^m \to \mathbb{A}^n$ and i_2 to a morphism $j_2 : \mathbb{A}^n \to \mathbb{A}^m$.

Let $\mathbf{x}$ be coordinates on $\mathbb{A}^n$ and $\mathbf{y}$ coordinates on $\mathbb{A}^m$. Then

$$(\mathbf{x}, \mathbf{y}) \mapsto (\mathbf{x}, \mathbf{y} + j_2(\mathbf{x}))$$

is an automorphism of $\mathbb{A}^{n+m}$, which sends the image of i_1' to

$$\mathrm{im}\big[i_1 \times i_2 : X \to \mathbb{A}^n \times \mathbb{A}^m\big].$$

Similarly,

$$(\mathbf{x}, \mathbf{y}) \mapsto (\mathbf{x} + j_1(\mathbf{y}), \mathbf{y})$$

is an automorphism of $\mathbb{A}^{n+m}$, which sends the image of i_2' to

$$\mathrm{im}\big[i_1 \times i_2 : X \to \mathbb{A}^n \times \mathbb{A}^m\big]. \quad \square$$

ASIDE 3.40. It is worthwhile to mention a local variant of (3.39). Let X be a scheme and $x \in X$ a point whose Zariski tangent space has dimension d. Then, for $m \geq 2d$, $x \in X$ has a unique embedding into a smooth scheme of dimension m, up to étale coordinate changes.

See [Jel87, Kal91] for affine versions.

LEMMA 3.41. *Let $h : Y \to X$ be a smooth morphism, $y \in Y$ a point and $i : X \hookrightarrow A$ a closed embedding. Then there are open neighborhoods $y \in Y^0 \subset Y$, $f(y) \in A_X^0 \subset A_X$, $X^0 = X \cap A_X^0$; a smooth morphism $h_A : A_Y^0 \to A_X^0$; and a closed embedding $j : Y^0 \hookrightarrow A_Y^0$ such that the following diagram is a fiber product square:*

$$\begin{array}{ccc}
Y^0 & \overset{j}{\hookrightarrow} & A_Y^0 \\
h \downarrow & \square & \downarrow h_A \\
X^0 & \overset{i}{\hookrightarrow} & A_X^0.
\end{array}$$

Proof. We prove this over infinite fields, which is the only case that we use.

The problem is local, and thus we may assume that X, Y, A_X are affine and $Y \subset X \times \mathbb{A}^N$. If h has relative dimension d, choose a general projection

$\sigma : \mathbb{A}_x^N \to \mathbb{A}_x^{d+1}$ such that $\sigma : h^{-1}(x) \to \mathbb{A}_x^{d+1}$ is finite and an embedding in a neighborhood of y. (Here we need that the residue field of x is infinite.) Thus, by shrinking Y, we may assume that Y is an open subset of a hypersurface $H \subset X \times \mathbb{A}^{d+1}$ and the first projection is smooth at $y \in H$. H is defined by an equation $\sum_I \phi_I z^I$, where the ϕ_I are regular functions on X and z denotes the coordinates on $\mathbb{A}^{d+1}$. Since $X \hookrightarrow A_X$ is a closed embedding, the ϕ_I extend to regular functions Φ_I on A_X. Set

$$A_Y := (\sum_I \Phi_I z^I = 0) \subset A_X \times \mathbb{A}^{d+1}.$$

Thus $Y \subset A_Y$ and the projection $A_Y \to A_X$ is smooth at y. Let $y \in A_Y^0 \subset A_Y$ and $A_X^0 \subset A_X$ be open sets such that the projection $h_A : A_Y^0 \to A_X^0$ is smooth and surjective. Set $Y^0 := Y \cap A_Y^0$.   □

The following comments on resolution for algebraic and analytic spaces are not used elsewhere in these notes.

3.42 (Algebraic spaces). All we need to know about algebraic spaces is that étale locally they are like schemes. That is, there is a (usually nonconnected) scheme of finite type U and an étale surjection $\sigma : U \to X$. We can even assume that U is affine.

The fiber product $V := U \times_X U$ is again a scheme of finite type with two surjective, étale projection morphisms $\rho_i : V \to U$, and for all purposes one can identify the algebraic space with the diagram of schemes

$$X = [\rho_1, \rho_2 : V \rightrightarrows U]. \tag{3.42.1}$$

The argument of (3.37) applies to show that any blow-up sequence functor $\mathcal{B}$ that is defined on affine schemes over a field k and commutes with étale surjections, has a unique extension to a blow-up sequence functor $\overline{\mathcal{B}}$, which is defined on all algebraic spaces over k. (See (3.105) for details.) Thus we obtain the following.

COROLLARY 3.43. *The theorems (3.35) and (3.36) also hold for algebraic spaces of finite type over a field of characteristic zero.*   □

3.44 (Analytic spaces). It was always understood that a good resolution method should also work for complex, real or p-adic analytic spaces. (See [GR71] for an introduction to analytic spaces.)

The traditional methods almost all worked well locally, but globalization sometimes presented technical difficulties. We leave it to the reader to follow the proofs in this chapter and see that they all extend to analytic spaces over locally compact fields, at least locally. Once, however, we have a locally defined blow-up sequence functor that commutes with smooth surjections, the argument of (3.37) shows that we get a globally defined blow-up sequence functor for small neighborhoods of compact sets on all analytic spaces. Once we have a resolution functor on neighborhoods of compact sets that commutes with open embeddings, we get resolution for any analytic space that is an increasing union of its compact subsets. Thus we obtain the following.

THEOREM 3.45. *Let K be a locally compact field of characteristic zero. There is a resolution functor $\mathcal{R} : X \to (\Pi_X : R(X) \to X)$ defined on all separable K-analytic spaces with the following properties.*

(1) $R(X)$ *is smooth.*
(2) $\Pi : R(X) \to X$ *is an isomorphism over the smooth locus X^{ns}.*
(3) $\Pi^{-1}(\operatorname{Sing} X)$ *is a divisor with simple normal crossing.*
(4) Π_X *is projective over any compact subset of X.*
(5) $\mathcal{R}$ *commutes with smooth K-morphisms.*   □

ASIDE 3.46. We give an example of a normal, proper surface S over $\mathbb{C}$ that cannot be embedded into a smooth scheme.

Start with $\mathbb{P}^1 \times C$, where C is any smooth curve of genus ≥ 1. Take two points $c_1, c_2 \in C$. Blow up $(0, c_1)$ and (∞, c_2) to get $f : T \to \mathbb{P}^1 \times C$. We claim the following.

(1) The birational transforms $C_1 \subset T$ of $\{0\} \times C$ and $C_2 \subset T$ of $\{\infty\} \times C$ can be contracted, and we get a normal, proper surface $g : T \to S$.
(2) If $\mathcal{O}_C(c_1)$ and $\mathcal{O}_C(c_2)$ are independent in $\operatorname{Pic}(C)$, then S can not be embedded into a smooth scheme.

To get the first part, it is easy to check that a multiple of the birational transform of $\{1\} \times C + \mathbb{P}^1 \times \{c_i\}$ on T is base point free and contracts C_i only, giving $g_i : T \to S_i$. Now $S_1 \setminus C_2$ and $S_2 \setminus C_1$ can be glued together to get $g : T \to S$.

If D is a Cartier divisor on S, then $\mathcal{O}_T(g^*D)$ is trivial on both C_1 and C_2. Therefore, $f_*(g^*D)$ is a Cartier divisor on $\mathbb{P}^1 \times C$ such that its restriction to $\{0\} \times C$ is linearly equivalent to a multiple of c_1 and its restriction to $\{\infty\} \times C$ is linearly equivalent to a multiple of c_2.

Since $\operatorname{Pic}(\mathbb{P}^1 \times C) = \operatorname{Pic}(C) \times \mathbb{Z}$ and $\mathcal{O}_C(c_1)$ and $\mathcal{O}_C(c_2)$ are independent in $\operatorname{Pic}(C)$, every Cartier divisor on S is linearly equivalent to a multiple of $\{1\} \times C$. Thus the points of $\{1\} \times C \subset S$ cannot be separated from each other by Cartier divisors on S.

Assume now that $S \hookrightarrow Y$ is an embedding into a smooth scheme. Pick a point $p \in \{1\} \times C \subset Y$, and let $p \in U \subset Y$ be an affine neighborhood. Any two points of U can be separated from each other by Cartier divisors on U. Since Y is smooth, the closure of a Cartier divisor on U is automatically Cartier on Y. Thus any two points of $U \cap S$ can be separated from each other by Cartier divisors on S, a contradiction.   □

An example of a toric variety with no Cartier divisors is given in [Ful93, p.65]. This again has no smooth embeddings.

3.4. Plan of the proof

This section contains a still somewhat informal review of the main steps of the proof. For simplicity, the role of the divisor E is ignored for now. All the definitions and theorems will be made precise later.

We need some way to measure how complicated an ideal sheaf is at a point. For the present proof a very crude measure—the order of vanishing or, simply, order—is enough.

DEFINITION 3.47. Let X be a smooth variety and $0 \neq I \subset \mathcal{O}_X$ an ideal sheaf. For a point $x \in X$ with ideal sheaf m_x, we define the *order of vanishing* or *order of I* at x to be

$$\mathrm{ord}_x I := \max\{r : m_x^r \mathcal{O}_{x,X} \supset I\mathcal{O}_{x,X}\}.$$

It is easy to see that $x \mapsto \mathrm{ord}_x I$ is a constructible and upper-semi-continuous function on X.

For an irreducible subvariety $Z \subset X$, we define the *order of I* along $Z \subset X$ as

$$\mathrm{ord}_Z I := \mathrm{ord}_\eta I, \quad \text{where } \eta \in Z \text{ is the generic point.}$$

Frequently we also use the notation $\mathrm{ord}_Z I = m$ (resp., $\mathrm{ord}_Z I \geq m$) when Z is not irreducible. In this case we always assume that the order of I at every generic point of Z is m (resp., $\geq m$).

The *maximal order* of I along $Z \subset X$ is

$$\text{max-ord}_Z I := \max\{\mathrm{ord}_z I : z \in Z\}.$$

We frequently use max-ord I to denote max-ord$_X I$.

If $I = (f)$ is a principal ideal, then the order of I at a point x is the same as the multiplicity of the hypersurface $(f = 0)$ at x. This is a simple but quite strong invariant.

In general, however, the order is a very stupid invariant. For resolution of singularities we always start with an embedding $X \hookrightarrow \mathbb{P}^N$, where N is larger than the embedding dimension of X at any point. Thus the ideal sheaf I_X of X contains an order 1 element at every point (the local equation of a smooth hypersurface containing X), so the order of I_X is 1 at every point of X. Hence the order of I_X does not "see" the singularities of X at all. (In the proof given in Section 3.12, trivial steps reduce the principalization of the ideal sheaf of $X \subset \mathbb{P}^N$ near a point $x \in X$ to the principalization of the ideal sheaf of $X \subset P$, where $P \subset \mathbb{P}^N$ is smooth and has the smallest possible dimension locally near x. Thus we start actual work only when ord $I \geq 2$.)

There is one useful property of $\mathrm{ord}_Z I$, which is exactly what we need: the number $\mathrm{ord}_Z I$ equals the multiplicity of $\pi^* I$ along the exceptional divisor of the blow-up $\pi : B_Z X \to X$.

DEFINITION 3.48 (Birational transform of ideals). Let X be a smooth variety and $I \subset \mathcal{O}_X$ an ideal sheaf. For $\dim X \geq 2$ an ideal cannot be written as the product of prime ideals, but the codimension 1 primes can be separated from the rest. That is, there is a unique largest effective divisor $\mathrm{Div}(I)$ such that $I \subset \mathcal{O}_X(-\mathrm{Div}(I))$, and we can write

$$I = \mathcal{O}_X\big(-\mathrm{Div}(I)\big) \cdot I_{\mathrm{cod} \geq 2}, \quad \text{where } \mathrm{codim}\,\mathrm{Supp}(\mathcal{O}_X/I_{\mathrm{cod} \geq 2}) \geq 2.$$

We call $\mathcal{O}_X\big(-\mathrm{Div}(I)\big)$ the *divisorial part* of I and $I_{\mathrm{cod} \geq 2} = \mathcal{O}_X\big(\mathrm{Div}(I)\big) \cdot I$ the *codimension ≥ 2 part* of I.

Let $f : X' \to X$ be a birational morphism between smooth varieties. Assume for simplicity that I has no divisorial part, that is, $I = I_{\mathrm{cod} \geq 2}$. We are interested in the codimension ≥ 2 part of f^*I, called the *birational transform* of I and denoted by $f_*^{-1}I$. (It is also frequently called the weak transform in the literature.) Thus

$$f_*^{-1}I = \mathcal{O}_{X'}\big(\mathrm{Div}(f^*I)\big) \cdot f^*I.$$

We have achieved principalization iff the codimension ≥ 2 part of f^*I is not there, that is, when $f_*^{-1}I = \mathcal{O}_{X'}$.

For reasons connected with (3.13.2), we also need another version, where we "pretend" that I has order m.

A *marked ideal sheaf* on X is a pair (I, m) where $I \subset \mathcal{O}_X$ is an ideal sheaf on X and m is a natural number.

Let $\pi : B_Z X \to X$ be the blow-up of a smooth subvariety Z and $E \subset B_Z X$ the exceptional divisor. Assume that $\mathrm{ord}_Z I \geq m$. Set

$$\pi_*^{-1}(I, m) := \big(\mathcal{O}_{B_Z X}(mE) \cdot \pi^*I, m\big),$$

and call it the *birational transform* of (I, m).

If $\mathrm{ord}_Z I = m$, then this coincides with $f_*^{-1}I$, but for $\mathrm{ord}_Z I > m$ the cosupport of $f_*^{-1}(I, m)$ also contains E. (We never use the case where $\mathrm{ord}_Z I < m$, since then $f_*^{-1}(I, m)$ is not an ideal sheaf.) One can iterate this procedure to define $f_*^{-1}(I, m)$ whenever $f : X' \to X$ is the composite of blow-ups of smooth irreducible subvarieties as above, but one has to be quite careful with this; see (3.63).

3.49 (Order reduction theorems). The technical core of the proof consists of two order reduction theorems using smooth blow-ups that match the order that we work with.

Let I be an ideal sheaf with $\mathrm{max\text{-}ord}\,I \leq m$. A *smooth blow-up sequence of order m* starting with (X, I) is a smooth blow-up sequence (3.29)

$$\Pi : (X_r, I_r) \xrightarrow{\pi_{r-1}} (X_{r-1}, I_{r-1}) \xrightarrow{\pi_{r-2}} \cdots \xrightarrow{\pi_1} (X_1, I_1) \xrightarrow{\pi_0} (X_0, I_0) = (X, I),$$

where each $\pi_i : X_{i+1} \to X_i$ is a smooth blow-up with center $Z_i \subset X_i$, the I_i are defined recursively by the formula $I_{i+1} := (\pi_i)_*^{-1} I_i$ and $\mathrm{ord}_{Z_i} I_i = m$ for every $i < r$.

A *blow-up sequence* of order $\geq m$ starting with a marked ideal (X, I, m) is defined analogously, except we use the recursion formula $(I_{i+1}, m) := (\pi_i)_*^{-1}(I_i, m)$ and we require $\text{ord}_{Z_i} I_i \geq m$ for every $i < r$.

Using these notions, the inductive versions of the main results are the following.

3.49.1 (Order reduction for ideals). *Let X be a smooth variety, $0 \neq I \subset \mathcal{O}_X$ an ideal sheaf and $m = \text{max-ord}\, I$. By a suitable blow-up sequence of order m we eventually get $f : X' \to X$ such that $\text{max-ord}\, f_*^{-1} I < m$.*

3.49.2 (Order reduction for marked ideals). *Let X be a smooth variety, $0 \neq I \subset \mathcal{O}_X$ an ideal sheaf and $m \leq \text{max-ord}\, I$ a natural number. By a suitable blow-up sequence of order $\geq m$, we eventually get $f : X' \to X$ such that $\text{max-ord}\, f_*^{-1}(I, m) < m$.*

We prove these theorems together in a spiraling induction with two main reduction steps.

$$\boxed{\begin{array}{c} \text{order reduction for marked ideals in dimension } n - 1 \\[4pt] \Downarrow \\[4pt] \text{order reduction for ideals in dimension } n \\[4pt] \Downarrow \\[4pt] \text{order reduction for marked ideals in dimension } n \end{array}}$$

The two steps are independent and use different methods.

The second implication is relatively easy and has been well understood for a long time. We leave it to Section 3.13.

Here we focus on the proof of the harder part, which is the first implication.

3.50 (The heart of the proof). Methods to deal with Problems (3.13.3–5) form the key steps of the proof. My approach is to break apart the traditional inductive proof. The problems can be solved independently but only for certain ideals. Then we need one more step to show that order reduction for an arbitrary ideal is equivalent to order reduction for an ideal with all the required good properties.

3.50.1 (Maximal contact). This deals with (3.13.3) by showing that for suitable hypersurfaces $H \subset X$ every step of an order reduction algorithm for (X, I) with $m = \text{max-ord}\, I$ is also a step of an order reduction algorithm for $(H, I|_H, m)$. This is explained in (3.51) and completed in Section 3.8.

3.50.2 (*D*-balanced ideals). Problem (3.13.4) has a solution for certain ideals only. For the so-called *D*-balanced ideals, the converse of maximal contact theory holds. That is, for every hypersurface $S \subset X$, every order reduction step for $(S, I|_S, m)$ is also an order reduction step for (X, I). This is outlined in (3.52) with all details in Section 3.9.

3.50.3 (MC-invariant ideals). The solution of (3.13.5) requires the consideration of MC-invariant ideals. For these, all hypersurfaces of maximal contact are locally analytically isomorphic, with an isomorphism preserving the ideal I. See (3.53), with full proofs in Section 3.10.

3.50.4 (Tuning of ideals). It remains to show that order reduction for an arbitrary ideal I is equivalent to order reduction for an ideal $W(I)$, which is both D-balanced and MC-invariant. This turns out to be surprisingly easy; see (3.54) and Section 3.11.

3.50.5 (Final assembly). The main remaining problem is that a hypersurface of maximal contact can be found only locally, not globally. The local pieces are united in Section 3.12, where we also take care of the divisor E, which we have ignored so far.

Let us now see these steps in more detail.

3.51 (Maximal contact). Following the examples (3.11) and (3.12), given X and I with $m = \text{max-ord}\, I$, we would like to find a smooth hypersurface $H \subset X$ such that order reduction for I follows from order reduction for $(I|_H, m)$.

As we noted in (3.13.3), first, we have to ensure that the points where the birational transform of I has order $\geq m$ stay on the birational transform of H all the time. That is, we want to achieve the following.

3.51.1 (Going-down property of maximal contact). Restriction (3.30.2.) from X to H gives an injection

$$
\boxed{\begin{array}{c}
\text{blow-up sequences of order } m \text{ for } (X, I) \\[4pt]
\cap \\[4pt]
\text{blow-up sequences of order } \geq m \text{ for } (H, I|_H, m)
\end{array}}
$$

If this holds, then we say that H is a hypersurface of *maximal contact*. At least locally these are easy to find using derivative ideals.

Derivations of a smooth variety X form a sheaf Der_X, locally generated by the usual partials $\partial/\partial x_i$. For an ideal sheaf I, let $D(I)$ denote the ideal sheaf generated by all derivatives of local sections of I. We can define higher-derivative ideals inductively by the rule $D^{i+1}(I) := D(D^i(I))$.

If $m = \text{max-ord}\, I$, we are especially interested in the largest nontrivial derivative ideal. It is also called the *ideal of maximal contacts*

$$
MC(I) := D^{m-1}(I) = \left(\frac{\partial^{m-1} f}{\partial x_1^{c_1} \cdots \partial x_n^{c_n}} : f \in I,\ \sum c_i = m - 1 \right).
$$

3.51.2 (Local construction of maximal contact). *For a point $p \in X$ with $m = \mathrm{ord}_p I$, let $h \in MC(I)$ be any local section with $\mathrm{ord}_p h = 1$. Then $H := (h = 0)$ is a hypersurface of maximal contact in an open neighborhood of p.*

In general, hypersurfaces of maximal contact do not exist globally, and they are not unique locally. We deal with these problems later.

3.52 (*D*-balanced ideals). It is harder to deal with (3.13.4). No matter how we choose the hypersurface of maximal contact H, sometimes the restriction $(I|_H, m)$ is "more singular" than I, in the sense that order reduction for $(I|_H, m)$ may involve blow-ups that are not needed for any order reduction procedure of I; see (3.82).

There are, however, some ideals for which this problem does not happen. To define these, we again need to consider derivatives.

If $\mathrm{ord}_p f = m$, then typically $\mathrm{ord}_p(\partial f/\partial x_i) = m - 1$, so a nontrivial ideal is never D-closed. The best one can hope for is that I is D-closed, after we "correct for the lowering of the order."

An ideal I with $m = \text{max-ord}\, I$ is called D-*balanced* if

$$\left(D^i(I)\right)^m \subset I^{m-i} \quad \forall\, i < m.$$

Such ideals behave very well with respect to restriction to smooth subvarieties and smooth blow ups.

3.52.1 (Going-up property of *D*-balanced ideals). Let I be a D-balanced ideal with $m = \text{max-ord}\, I$. Then for any smooth hypersurface $S \subset X$ such that $S \not\subset \mathrm{cosupp}\, I$, push-forward (3.30.3.) from S to X gives an injection

$$
\boxed{
\begin{array}{c}
\text{blow-up sequences of order } m \text{ for } (X, I) \\[4pt]
\cup \\[4pt]
\text{blow-up sequences of order } \geq m \text{ for } (S, I|_S, m)
\end{array}
}
$$

Example 3.52.2. Start with the double point ideal $I = (xy - z^n)$. Restricting to $S = (x = 0)$ creates an n-fold line, and blowing up this line is not an order 2 blow-up for I.

We can see that

$$I + D(I)^2 = (xy, x^2, y^2, xz^{n-1}, yz^{n-1}, z^n)$$

is D-balanced. If we restrict $I + D(I)^2$ to $(x = 0)$, we get the ideal (y^2, yz^{n-1}, z^n). It is easy to check that the whole resolution of S is correctly predicted by order reduction for (y^2, yz^{n-1}, z^n).

Putting (3.51.1) and (3.52.1) together, we get the first dimension reduction result.

Corollary 3.52.3. (Maximal contact for D-balanced ideals). *Let I be a D-balanced ideal with $m = \text{max-ord}\, I$ and $H \subset X$ a smooth hypersurface of maximal contact. Then we have an equivalence*

$$
\boxed{
\begin{array}{c}
\textit{blow-up sequences of order } m \textit{ for } (X, I) \\
\| \\
\textit{blow-up sequences of order} \geq m \textit{ for } (H, I|_H, m)
\end{array}
}
$$

This equivalence suggests that the choice of H should not be important at all. However, in order to ensure functoriality we have to choose a particular resolution. Thus we still need to show that our particular choices are independent of H. A truly "canonical" resolution process would probably take care of such problems automatically, but it seems that one has to make at least some artificial choices.

3.53 (MC-invariant ideals). Dealing with (3.13.5) is again possible only for certain ideals.

We say that I is *maximal contact invariant* or *MC-invariant* if

$$MC(I) \cdot D(I) \subset I. \tag{3.53.1}$$

Written in the equivalent form

$$D^{m-1}(I) \cdot D(I) \subset I, \tag{3.53.2}$$

it is quite close in spirit to the D-balanced condition. The expected order of $D^{m-1}(I) \cdot D(I)$ is m, so it is sensible to require inclusion. There is no need to correct for the change of order first.

For MC-invariant ideals the hypersurfaces of maximal contact are still not unique, but different choices are equivalent under local analytic isomorphisms (3.55).

3.53.3 (Formal uniqueness of maximal contact). *Let I be an MC-invariant ideal sheaf on X and $H_1, H_2 \subset X$ two hypersurfaces of maximal contact through a point $x \in X$. Then there is a local analytic automorphism (3.55) $\phi : (x \in \hat{X}) \to (x \in \hat{X})$ such that*

  (i) $\phi^{-1}(\hat{H}_1) = \hat{H}_2$,
  (ii) $\phi^* \hat{I} = \hat{I}$, *and*
  (iii) ϕ *is the identity on* $\text{cosupp}\, \hat{I}$.

3.54 (Tuning of ideals). Order reduction using dimension induction is now in quite good shape for ideals that are both D-balanced and MC-invariant.

The rest is taken care of by "tuning" the ideal I first. (I do not plan to give a precise meaning to the word "tuning." The terminology follows

[Włous05]. The notion of tuning used in [EH02] is quite different.) There are many ways to tune an ideal; here is one of the simplest ones.

To an ideal I of order m, we would like to associate the ideal generated by all products of derivatives of order at least m. The problem with this is that if f has order m, then $\partial f / \partial x_i$ has order $m - 1$, and so we are able to add $(\partial f / \partial x_i)^2$ (which has order $2m - 2$), but we really would like to add $(\partial f / \partial x_i)^{m/(m-1)}$ (which should have order m in any reasonable definition).

We can avoid these fractional exponent problems by working with all products of derivatives whose order is sufficiently divisible. For instance, the condition (order) $\geq m!$ works.

Enriching an ideal with its derivatives was used by Hironaka [Hir77] and then developed by Villamayor [Vil89]. A larger ideal is introduced in [Włous05]. The ideal $W(I)$ introduced below is even larger, and this largest choice seems more natural to me. That is, we set

$$W(I) := \left(\prod_{j=0}^{m} \left(D^j(I) \right)^{c_j} : \sum (m - j)c_j \geq m! \right) \subset \mathcal{O}_X. \qquad (3.54.1)$$

The ideal $W(I)$ has all the properties that we need.

Theorem 3.54.2. (Well-tuned ideals). Let X be a smooth variety, $0 \neq I \subset \mathcal{O}_X$ an ideal sheaf and $m = \max\text{-ord}\, I$. Then

    (i) $\max\text{-ord}\, W(I) = m!$,
    (ii) $W(I)$ is D-balanced,
    (iii) $W(I)$ is MC-invariant, and
    (iv) there is an equivalence

> blow-up sequences of order m for (X, I)
>
> $\|$
>
> blow-up sequences of order $m!$ for $(X, W(I))$

3.54.3. It should be emphasized that there are many different ways to choose an ideal with the properties of $W(I)$ as above, but all known choices have rather high order.

I chose the order $m!$ for notational simplicity, but one could work with any multiple of $\mathrm{lcm}(1, 2, \ldots, m)$ instead. The smallest choice would be $\mathrm{lcm}(1, 2, \ldots, m)$, which is roughly like e^m. As discussed in (3.7.2), this is still too big for effective computations. Even if we fix the order to be $m!$, many choices remain.

3.54.4. Similar constructions are also considered by Kawanoue [Kaw06] and by Villamayor [Vil06].

DEFINITION 3.55 (Completions). This is the only piece of commutative algebra that we use.

For a local ring (R, m) its *completion* in the m-adic topology is denoted by $\hat{R}$; cf. [AM69, Chap.10]. If X is a k-variety and $x \in X$, then we denote by $\hat{X}_x$ or by $\hat{X}$ the completion of X at x, which is $\operatorname{Spec}_k \hat{\mathcal{O}}_{x,X}$.

We say that $x \in X$ and $y \in Y$ are *formally* isomorphic if $\hat{X}_x$ is isomorphic to $\hat{Y}_y$.

We need Krull's intersection theorem (cf. [AM69, 10.17]), which says that for an ideal I in a Noetherian local ring (R, m) we have

$$I = \cap_{s=1}^{\infty}(I + m^s).$$

In geometric language this implies that if $Z, W \subset X$ are two subschemes such that $\hat{Z}_x = \hat{W}_x$, then there is an open neighborhood $x \in U \subset X$ such that $Z \cap U = W \cap U$.

If $p \in X$ is closed, then $\hat{\mathcal{O}}_{p,X} \cong k(p)[[x_1, \ldots, x_n]]$, where $x_1, \ldots, x_n$ are local coordinates. If $k(p) = k$ or, more generally, when there is a *field of representatives* (that is, a subfield $k' \subset \hat{\mathcal{O}}_{p,X}$ isomorphic to $k(p)$), this is proved in [Sha94, II.2]. In characteristic zero one can find k' as follows. The finite field extension $k(p)/k$ is generated by a simple root of a polynomial $f(y) \in k[y] \subset \hat{\mathcal{O}}_{p,X}[y]$. Modulo the maximal ideal, $f(y)$ has a linear factor by assumption, and thus by the general Hensel lemma (1.92), $f(y)$ has a linear factor and hence a root $\alpha \in \hat{\mathcal{O}}_{p,X}$. Then $k' = k(\alpha)$ is the required subfield. (Note that usually one cannot find such $k' \subset \mathcal{O}_{p,X}$, and so the completion is necessary.)

REMARK 3.56. By the approximation theorem of [Art69], $x \in X$ and $y \in Y$ are formally isomorphic iff there is a $z \in Z$ and étale morphisms

$$(x \in X) \leftarrow (z \in Z) \rightarrow (y \in Y).$$

This implies that any resolution functor that commutes with étale morphisms also commutes with formal isomorphisms.

Our methods give resolution functors that commute with formal isomorphisms by construction, so we do not need to rely on [Art69].

ASIDE 3.57 (Maximal contact in positive characteristic). Maximal contact, in the form presented above, works in positive characteristic as long as the order of the ideal is less than the characteristic but fails in general. In some cases there is no smooth hypersurface at all that contains the set of points where the order is maximal. The following example is taken from [Nar83]. In characteristic 2 consider

$$X := (x^2 + yz^3 + zw^3 + y^7w = 0) \subset \mathbb{A}^4.$$

The maximal multiplicity is 2, and the singular locus is given by

$$x^2 + yz^3 + zw^3 + y^7w = z^3 + y^6w = yz^2 + w^3 = zw^2 + y^7 = 0.$$

It contains the monomial curve

$$C := \text{im}[t \mapsto (t^{32}, t^7, t^{19}, t^{15})]$$

(in fact, it is equal to it). C is not contained in any smooth hypersurface. Indeed, assume that $(F = 0)$ is a hypersurface containing C that is smooth at the origin. Then one of x, y, z, w appears linearly in F and $F(t^{32}, t^7, t^{19}, t^{15}) \equiv 0$. The linear term gives a nonzero t^m for some $m \in \{32, 7, 19, 15\}$, which must be canceled by another term t^m. Thus we can write $m = 32a + 7b + 19c + 15d$, where $a + b + c + d \geq 2$ and $a, b, c, d \geq 0$. This is, however, impossible since none of the numbers $32, 7, 19, 15$ is a positive linear combination of the other three.

3.5. Birational transforms and marked ideals

3.58 (Birational transform of ideals). Let X be a smooth scheme over a field k, $Z \subset X$ a smooth subscheme and $\pi : B_Z X \to X$ the blow-up with exceptional divisor $F \subset B_Z X$. Let $Z = \cup Z_j$ and $F = \cup F_j$ be the irreducible components.

Let $I \subset \mathcal{O}_X$ be an ideal sheaf, and set $\text{ord}_{Z_j} I = m_j$. Then $\pi^* I \subset \mathcal{O}_{B_Z X}$ vanishes along F_j with multiplicity m_j, and we aim to remove the ideal sheaf $\mathcal{O}_{B_Z X}(-\sum m_j F_j)$ from $\pi^* I$. That is, define the *birational transform* (also called the controlled transform or weak transform in the literature) of I by the formula

$$\pi_*^{-1} I := \mathcal{O}_{B_Z X}(\textstyle\sum_j m_j F_j) \cdot \pi^* I \subset \mathcal{O}_{B_Z X}. \qquad (3.58.1)$$

This is consistent with the definition given in (3.49) for the case $I = I_{\text{cod} \geq 2}$.

Warning. If $Z \subset X$ is a smooth divisor, then the blow-up is trivial. Hence $\pi_{Z,X} : B_Z X \cong X$ is the identity map, and

$$\pi_*^{-1} I := \mathcal{O}_X(\textstyle\sum_j m_j Z_j) \cdot I$$

depends not only on $\pi = \pi_{Z,X}$ but also on the center Z of the blow-up. Unfortunately, I did not find any good way to fix this notational inconsistency.

One problem we have to deal with in resolutions is that if $Z \subset H \subset X$ is a smooth hypersurface with birational transform $B_Z H \subset B_Z X$ and projection $\pi_H : B_Z H \to H$, then restriction to H does not commute with taking birational transform. That is,

$$(\pi_H)_*^{-1}(I|_H) \supset (\pi_*^{-1} I)|_{B_Z H}, \qquad (3.58.2)$$

but equality holds only if $\text{ord}_Z I = \text{ord}_Z(I|_H)$.

The next definition is designed to remedy this problem. We replace the ideal sheaf I by a pair (I, m), where m keeps track of the order of vanishing that we pretend to have. The advantage is that we can redefine the notion of birational transform to achieve equality in (3.58.2).

DEFINITION 3.59. Let X be a smooth scheme. A *marked function* on X is a pair (f, m), where f is a regular function on (some open set of) X and m a natural number.

A *marked ideal sheaf* on X is a pair (I, m), where $I \subset \mathcal{O}_X$ is an ideal sheaf on X and m a natural number.

The *cosupport* of (I, m) is defined by

$$\mathrm{cosupp}(I, m) := \{x \in X : \mathrm{ord}_x I \geq m\}.$$

The *product* of marked functions or marked ideal sheaves is defined by

$$(f_1, m_1) \cdot (f_2, m_2) := (f_1 f_2, m_1 + m_2), \quad (I_1, m_1) \cdot (I_2, m_2) := (I_1 I_2, m_1 + m_2).$$

The *sum* of marked functions or marked ideal sheaves is only sensible when the markings are the same:

$$(f_1, m) + (f_2, m) := (f_1 + f_2, m) \quad \text{and} \quad (I_1, m) + (I_2, m) := (I_1 + I_2, m).$$

The cosupport has the following elementary properties:

(1) if $I \subset J$ then $\mathrm{cosupp}(I, m) \supset \mathrm{cosupp}(J, m)$,
(2) $\mathrm{cosupp}(I_1 I_2, m_1 + m_2) \supset \mathrm{cosupp}(I_1, m_1) \cap \mathrm{cosupp}(I_2, m_2)$,
(3) $\mathrm{cosupp}(I, m) = \mathrm{cosupp}(I^c, mc)$,
(4) $\mathrm{cosupp}(I_1 + I_2, m) = \mathrm{cosupp}(I_1, m) \cap \mathrm{cosupp}(I_2, m)$.

DEFINITION 3.60. Let X be a smooth variety, $Z \subset X$ a smooth subvariety and $\pi : B_Z X \to X$ the blow-up with exceptional divisor $F \subset B_Z X$. Let (I, m) be a marked ideal sheaf on X such that $m \leq \mathrm{ord}_Z I$. In analogy with (3.58) we define the *birational transform* of (I, m) by the formula

$$\pi_*^{-1}(I, m) := \big(\mathcal{O}_{B_Z X}(mF) \cdot \pi^* I, m \big). \tag{3.60.1}$$

Informally speaking, we use the definition (3.58.2), but we "pretend that $\mathrm{ord}_Z I = m$."

As in (3.58), it is worth calling special attention to the case where Z has codimension 1 in X. Then $B_Z X \cong X$, and so scheme-theoretically there is no change. However, the vanishing order of $\pi_*^{-1}(I, m)$ along Z is m less than the vanishing order of I along Z.

In order to do computations, choose local coordinates $(x_1, \ldots, x_n)$ such that $Z = (x_1 = \cdots = x_r = 0)$. Then

$$y_1 = \tfrac{x_1}{x_r}, \ldots, y_{r-1} = \tfrac{x_{r-1}}{x_r}, \ y_r = x_r, \ldots, y_n = x_n \tag{3.60.2}$$

give local coordinates on a chart of $B_Z X$, and we define

$$\pi_*^{-1}(f, m) := \big(y_r^{-m} f(y_1 y_r, \ldots, y_{r-1} y_r, y_r, \ldots, y_n), m \big). \tag{3.60.3}$$

This formula is the one we use to compute with blow-ups, but it is coordinate system dependent. As we change coordinates, the result of π_*^{-1} changes by a unit. So we are free to use π_*^{-1} to compute the birational transform of ideal sheaves, but one should not use it for individual functions, whose birational transform cannot be defined (as a function).

The following lemmas are easy.

LEMMA 3.61. *Let X be a smooth variety, $Z \subset X$ a smooth subvariety, $\pi : B_Z X \to X$ the blow-up and $I \subset \mathcal{O}_X$ an ideal sheaf. Assume that $\mathrm{ord}_Z I = \text{max-ord}\, I$. Then*

$$\text{max-ord}\, \pi_*^{-1} I \leq \text{max-ord}\, I.$$

Proof. Choose local coordinates as above, and pick $f(x_1, \ldots, x_n) \in I$ such that $\mathrm{ord}_p f = \text{max-ord}\, I = m$. Its birational transform is computed as

$$\pi_*^{-1} f = y_r^{-m} f(y_1 y_r, \ldots, y_{r-1} y_r, y_r, \ldots, y_n).$$

Since $f(x_1, \ldots, x_n)$ contains a monomial of degree m, the corresponding monomial in $f(y_1 y_r, \ldots, y_{r-1} y_r, y_r, \ldots, y_n)$ has degree $\leq 2m$, and thus in $\pi_*^{-1} f$ we get a monomial of degree $\leq 2m - m = m$.

This shows that $\mathrm{ord}_{p'} \pi_*^{-1} I \leq m$, where $p' \in B_Z X$ denotes the origin of the chart we consider. Performing a linear change of the $(x_1, \ldots, x_r)$-coordinates moves the origin of the chart, and every preimage of p appears as the origin after a suitable linear change. Thus our computation applies to all points of the exceptional divisor of $B_Z X$. □

LEMMA 3.62. *Let the notation be as in (3.61). Let $Z \subsetneq H \subset X$ be a smooth hypersurface with birational transform $B_Z H \subset B_Z X$ and projection $\pi_H : B_Z H \to H$. If $m \leq \mathrm{ord}_Z I$ and $I|_H \neq 0$, then*

$$(\pi_H)_*^{-1}(I|_H, m) = \big(\pi_*^{-1}(I, m)\big)|_{B_Z H}.$$

Proof. Again choose coordinates and assume that $H = (x_1 = 0)$. Working with the chart as in (3.60.2), the birational transform of H is $(y_1 = 0)$, and we see that it does not matter whether we set first $x_1 = 0$ and compute the transform or first compute the transform and then set $y_1 = 0$. We still need to contemplate what happens in the chart

$$z_1 = x_1, z_2 = \frac{x_2}{x_1}, \ldots, z_r = \frac{x_r}{x_1}, z_{r+1} = x_{r+1}, \ldots, z_n = x_n.$$

This chart, however, does not contain any point of the birational transform of H, so it does not matter. □

Note that (3.62) fails if $Z = H$. In this case $I|_H$ is the zero ideal, π_Z is an isomorphism, and we have only the bad chart, which we did not need to consider in the proof above. Because of this, we will have to consider codimension 1 subsets of $\mathrm{cosupp}\, I$ separately.

WARNING 3.63. Note that, while the birational transform of an ideal with $I = I_{\mathrm{cod} \geq 2}$ is defined for an arbitrary birational morphism (3.58), we have defined the birational transform of a marked ideal only for a single smooth blow-up (3.60). This can be extended to a sequence of smooth blow-ups, but one has to be very careful. Let

$$\Pi : X' = X_r \xrightarrow{\pi_{r-1}} X_{r-1} \xrightarrow{\pi_{r-2}} \cdots \xrightarrow{\pi_1} X_1 \xrightarrow{\pi_0} X_0 = X \qquad (3.63.1)$$

be a smooth blow-up sequence. We can inductively define the birational transforms of the marked ideal (I, m) by

(1) $(I_0, m) := (I, m)$, and
(2) $(I_{j+1}, m) := (\pi_j)_*^{-1}(I_j, m)$ as in (3.60).

At the end we get (I_r, m), which I rather sloppily also denote by $\Pi_*^{-1}(I, m)$.

It is very important to keep in mind that this notation assumes that we have a particular blow-up sequence in mind. That is, $\Pi_*^{-1}(I, m)$ depends not only on the morphism Π but on the actual sequence of blow-ups we use to get it.

Consider, for instance, the blow-ups

$$\Pi : X_2 \xrightarrow{\pi_1} X_1 = B_p X_0 \xrightarrow{\pi_0} X_0,$$
$$\| \qquad\qquad \| $$
$$\Sigma : X_2' \xrightarrow{\sigma_1} X_1' = B_C X_0 \xrightarrow{\sigma_0} X_0$$

introduced in (3.33).

Let us compute the birational transforms of $(I, 1)$, where $I := I_C$. The first blow-up sequence gives

$$(\pi_0)_*^{-1}(I, 1) = \mathcal{O}_{X_1}(E_0) \cdot \pi_0^* I \quad \text{and}$$
$$(\pi_1)_*^{-1}\big((\pi_0)_*^{-1}(I, 1)\big) = \mathcal{O}_{X_2}(E_1) \cdot \pi_1^*\big((\pi_0)_*^{-1}(I, 1)\big)$$
$$= \mathcal{O}_{X_2}(E_0 + E_1) \cdot \Pi^* I.$$

On the other hand, the second blow-up sequence gives

$$(\sigma_0)_*^{-1}(I, 1) = \mathcal{O}_{X_1'}(E_0') \cdot \sigma_0^* I \quad \text{and}$$
$$(\sigma_1)_*^{-1}\big((\sigma_0)_*^{-1}(I, 1)\big) = \mathcal{O}_{X_2'}(E_1') \cdot \sigma_1^*\big((\sigma_0)_*^{-1}(I, 1)\big)$$
$$= \mathcal{O}_{X_2'}(E_0' + 2E_1') \cdot \Sigma^* I$$

since $\sigma_1^* \mathcal{O}_{X_1'}(E_0') = \mathcal{O}_{X_2'}(E_0' + E_1')$.

Thus $\Pi_*^{-1}(I, 1) \neq \Sigma_*^{-1}(I, 1)$, although $\Pi = \Sigma$.

3.6. The inductive setup of the proof

In this section we set up the final notation and state the main order reduction theorems.

NOTATION 3.64. For the rest of the chapter, (X, I, E) or (X, I, m, E) denotes a *triple*[1], where

(1) X is a smooth, equidimensional (possibly reducible) scheme of finite type over a field of characteristic zero,
(2) $I \subset \mathcal{O}_X$ (resp., (I, m)) is a coherent ideal sheaf (resp., coherent marked ideal sheaf), which is nonzero on every irreducible component of X, and

---

[1] I consider the pair (I, m) as one item, so (X, I, m, E) is still a triple.

(3) $E = (E^1, \ldots, E^s)$ is an ordered set of smooth divisors on X such that $\sum E^i$ is a simple normal crossing divisor. Each E^i is allowed to be reducible or empty.

The divisor E plays an ancillary role as a device that keeps track of the exceptional divisors that we created and of the order in which we created them. As we saw in (3.6.3), one has to carry along some information about the resolution process.

As we observed in (3.6.2) and (3.28.1), it is necessary to blow up disjoint subvarieties simultaneously. Thus we usually do get reducible smooth divisors E^j.

DEFINITION 3.65. Given (X, I, E) with max-ord $I = m$, a *smooth blow-up* of order m is a smooth blow-up $\pi : B_Z X \to X$ with *center* Z such that

(1) $Z \subset X$ has simple normal crossings only with E, and
(2) $\mathrm{ord}_Z I = m$.

The *birational transform* of (X, I, E) under the above blow-up is

$$\pi_*^{-1}(X, I, E) = \left(B_Z X, \pi_*^{-1} I, \pi_{\mathrm{tot}}^{-1}(E) \right).$$

Here $\pi_*^{-1} I$ is the birational transform of I as defined in (3.58), and $\pi_{\mathrm{tot}}^{-1}(E)$ consists of the birational transform (2.1) of E (with the same ordering as before) plus the exceptional divisor $F \subset B_Z X$ added as the last divisor. It is called the *total transform* of E. (If π is a trivial blow-up, then $\pi_{\mathrm{tot}}^{-1}(E) = E + Z$.)

A *smooth blow-up* of (X, I, m, E) is a smooth blow-up $\pi : B_Z X \to X$ such that

(1') $Z \subset X$ has simple normal crossings only with E, and
(2') $\mathrm{ord}_Z I \geq m$.

The *birational transform* of (X, I, m, E) under the above blow-up is defined as

$$\pi_*^{-1}(X, I, m, E) = \left(B_Z X, \pi_*^{-1}(I, m), \pi_{\mathrm{tot}}^{-1}(E) \right).$$

DEFINITION 3.66. A *smooth blow-up sequence* of order m and of length r starting with (X, I, E) such that max-ord $I = m$ is a smooth blow-up sequence (3.30)

$$\Pi : (X_r, I_r, E_r) \xrightarrow{\pi_{r-1}} (X_{r-1}, I_{r-1}, E_{r-1}) \xrightarrow{\pi_{r-2}} \cdots$$
$$\xrightarrow{\pi_1} (X_1, I_1, E_1) \xrightarrow{\pi_0} (X_0, I_0, E_0) = (X, I, E),$$

where

(1) the (X_i, I_i, E_i) are defined recursively by the formula

$$(X_{i+1}, I_{i+1}, E_{i+1}) := \left(B_{Z_i} X_i, (\pi_i)_*^{-1} I_i, (\pi_i)_{\mathrm{tot}}^{-1} E_i \right),$$

(2) each $\pi_i : X_{i+1} \to X_i$ is a smooth blow-up with center $Z_i \subset X_i$ and exceptional divisor $F_{i+1} \subset X_{i+1}$,

(3) for every i, $Z_i \subset X_i$ has simple normal crossings with E_i, and

(4) for every i, $\mathrm{ord}_{Z_i} I_i = m$.

Similarly, a *smooth blow-up sequence* of order $\geq m$ and of length r starting with (X, I, m, E) is a smooth blow-up sequence

$$\Pi : (X_r, I_r, m, E_r) \xrightarrow{\pi_{r-1}} (X_{r-1}, I_{r-1}, m, E_{r-1}) \xrightarrow{\pi_{r-2}} \cdots$$
$$\xrightarrow{\pi_1} (X_1, I_1, m, E_1) \xrightarrow{\pi_0} (X_0, I_0, m, E_0) = (X, I, m, E),$$

where

(1') the (X_i, I_i, m, E_i) are defined recursively by the formula

$$(X_{i+1}, I_{i+1}, m, E_{i+1}) := \left(B_{Z_i} X_i, (\pi_i)_*^{-1}(I_i, m), (\pi_i)_{\mathrm{tot}}^{-1} E_i \right),$$

(2'–3') the sequence satisfies (2) and (3) above, and

(4') for every i, $\mathrm{ord}_{Z_i} I_i \geq m$.

As we noted in (3.60), we allow the case where $Z_i \subset X_i$ has codimension 1. In this case π_{i+1} is an isomorphism, but $I_{i+1} \neq I_i$.

We also use the notation

$$\begin{aligned} \Pi_*^{-1}(X, I, E) &:= \left(X_r, \Pi_*^{-1} I, \Pi_{\mathrm{tot}}^{-1}(E) \right) \\ &:= (X_r, I_r, E_r), \end{aligned}$$

but keep in mind that, as we saw in (3.63), this depends on the whole blow-up sequence and not only on Π.

We also enrich the definition of *blow-up sequence functors* considered in (3.31). From now on, we consider functors $\mathcal{B}$ such that $\mathcal{B}(X, I, E)$ (resp., $\mathcal{B}(X, I, m, E)$) is a blow-up sequence starting with (X, I, E) (resp., (X, I, m, E)) as above. That is, from now on we consider the sheaves I_i and the divisors E_i as part of the functor. Since these are uniquely determined by (X, I, E) and the blow-ups π_i, this is a minor notational change.

REMARK 3.67. The difference between the marked and unmarked versions is significant, since the birational transforms of the ideals are computed differently.

There is one case, however, when one can freely pass between the two versions. If I is an ideal with max-ord $I = m$, then in any blow-up sequence of order $\geq m$ starting with (X, I, m, E), max-ord $I_i \leq m$ by (3.61), and so every blow-up has order $= m$. Thus, by deleting m, we automatically get a blow-up sequence of order m starting with (X, I, E). The converse also holds.

We can now state the two main technical theorems that combine to give an inductive proof of resolution.

THEOREM 3.68 (Order reduction for ideals). *For every m there is a smooth blow-up sequence functor $\mathcal{BO}_m$ of order m (3.31) that is defined on* *triples* (X, I, E) *with* max-ord $I \leq m$ *such that if* $\mathcal{BO}_m(X, I, E) =$

$$\Pi : (X_r, I_r, E_r) \xrightarrow{\pi_{r-1}} (X_{r-1}, I_{r-1}, E_{r-1}) \xrightarrow{\pi_{r-2}} \cdots$$
$$\xrightarrow{\pi_1} (X_1, I_1, E_1) \xrightarrow{\pi_0} (X_0, I_0, E_0) = (X, I, E),$$

then

(1) max-ord $I_r < m$, *and*
(2) $\mathcal{BO}_m$ *commutes with smooth morphisms (3.34.1) and with change of fields (3.34.2).*

In our examples, the case max-ord $I < m$ is trivial, that is, $X_r = X$.

THEOREM 3.69 (Order reduction for marked ideals). *For every m there is a smooth blow-up sequence functor $\mathcal{BMO}_m$ of order $\geq m$ (3.31) that is defined on triples (X, I, m, E) such that if $\mathcal{BMO}_m(X, I, m, E) =$*

$$\Pi : (X_r, I_r, m, E_r) \xrightarrow{\pi_{r-1}} (X_{r-1}, I_{r-1}, m, E_{r-1}) \xrightarrow{\pi_{r-2}} \cdots$$
$$\xrightarrow{\pi_1} (X_1, I_1, m, E_1) \xrightarrow{\pi_0} (X_0, I_0, m, E_0),$$

then

(1) max-ord $I_r < m$, *and*
(2) $\mathcal{BMO}_m$ *commutes with smooth morphisms (3.34.1) and also with change of fields (3.34.2).*

3.70 (Main inductive steps of the proof). We prove (3.68) and (3.69) together in two main reduction steps.

(3.69) in dimensions $\leq n - 1$
$\Downarrow$ (3.70.1)
(3.68) in dimension n
$\Downarrow$ (3.70.2)
(3.69) in dimension n

The easier part is (3.70.2). Its proof is given in Section 3.13. Everything before that is devoted to proving (3.70.1).

We can start the induction with the case $\dim X = 0$. Here $I = \mathcal{O}_X$ since I is assumed nonzero on every irreducible component of X. Everything is resolved without blow-ups.

The case $\dim X = 1$ is also uninteresting. The cosupport of an ideal sheaf is a Cartier divisor and our algorithm tells us to blow up $Z := $ cosupp(I, m). In the unmarked case $m = $ max-ord I. After one blow-up I is replaced by $I' := I \otimes \mathcal{O}_X(Z)$ which has order $< m$. In the marked case max-ord$\left(I \otimes \mathcal{O}_X(Z)\right) < $ max-ord I. Thus, after finitely many steps, the maximal order drops below m.

The 2-dimensional case is quite a bit more involved since it includes the resolution of plane curve singularities (essentially as in Section 1.10) and the principalization of ideal sheaves studied in Section 1.9.

3.71 (From principalization to resolution). As we saw in the proof of (3.27), one can prove the existence of resolutions for quasi-projective schemes using (3.68), but it is not clear that the resolution is independent of the projective embedding chosen. In order to prove it, we establish two additional properties of the functors $\mathcal{BO}$ and $\mathcal{BMO}$.

Claim 3.71.1. $\mathcal{BMO}_m(X, I, m, \emptyset) = \mathcal{BO}_m(X, I, \emptyset)$ if $m = \text{max-ord}\, I$.

Claim 3.71.2. Let $\tau : Y \hookrightarrow X$ be a closed embedding of smooth schemes and $J \subset \mathcal{O}_Y$ and $I \subset \mathcal{O}_X$ ideal sheaves such that J is nonzero on every irreducible component of Y and $\tau_*(\mathcal{O}_Y/J) = \mathcal{O}_X/I$. Then

$$\mathcal{BMO}_1(X, I, 1, \emptyset) = \tau_* \mathcal{BMO}_1(Y, J, 1, \emptyset).$$

In both of these claims we assume that $E = \emptyset$. One can easily extend (3.71.1) to arbitrary E, by slightly changing the definition (3.110). The situation with (3.71.2) is more problematic. If $E \neq \emptyset$, then (3.71.2) fails in some cases when cosupp J contains some irreducible components of $Y \cap E$. Most likely, this can also be fixed with relatively minor changes, but I do not know how.

Note also that (3.71.2) would not make sense for any marking different from $m = 1$. Indeed, the ideal I contains the local equations of Y, thus it has order 1. Thus $\mathcal{BMO}_m(X, I, m, \emptyset)$ is the identity for any $m \geq 2$.

3.72 (Proof of (3.69) & (3.71.2) $\Rightarrow$ (3.35)). The only tricky point is that in (3.35) E is a usual divisor but (3.68) assumes that the index set of E is ordered. We can order the index set somehow, so the existence of a principalization is not a problem. However, if we want functoriality, then we should not introduce arbitrary choices in the process.

If, by chance, the irreducible components of E are disjoint, then we can just declare that E is a single divisor, since in (3.68) we allow the components of E to be reducible. Next we show how to achieve this by some preliminary blow-ups.

Let $E_1, \ldots, E_k$ be the irreducible components of E. We make the E_i disjoint in $k - 1$ steps.

First, let $Z_0 \subset X_0 = X$ be the subset where all of the $E_1, \ldots, E_k$ intersect. Let $\pi_0 : X_1 \to X_0$ be the blow-up of Z_0 with exceptional divisor F^1. Note that the $(\pi_0)_*^{-1} E_1, \ldots, (\pi_0)_*^{-1} E_k$ do not have any k-fold intersections.

Next let $Z_1 \subset X_1$ be the subset where $k - 1$ of the $(\pi_0)_*^{-1} E_i$ intersect. Z_1 is smooth since the $(\pi_0)_*^{-1} E_i$ do not have any k-fold intersections. Let $\pi_1 : X_2 \to X_1$ be the blow-up of Z_1 with exceptional divisor F^2. Note that the $(\pi_0 \pi_1)_*^{-1} E_i$ do not have any $(k-1)$-fold intersections.

Next let $Z_2 \subset X_2$ be the subset where $k-2$ of the $(\pi_0 \pi_1)_*^{-1} E_i$ intersect, and so on.

After $(k-1)$-steps we get rid of all pairwise intersections as well. The end result is $\pi : X' \to X$ such that $E^0 := \pi_*^{-1}(E_1 + \cdots + E_k)$ is a smooth divisor. Let $E^1, \ldots, E^{k-1}$ denote the birational transforms of $F^1, \ldots, F^{k-1}$.

Thus $(X', \pi^*I, \sum_{i=0}^{k-1} E^i)$ satisfies the assumptions of (3.68).

(It may seem natural to start with $\dim X$-fold intersections instead of k-fold intersections. We want functoriality with respect to all smooth morphisms, so we should not use the dimension of X in constructing the resolution process. However, ultimately the difference is only in some empty blow ups, and we can forget about those at the end.)

The rest is straightforward. Construct

$$\Pi_{(X,I,E)} : X_r \to \cdots \to X_s = X' \to \cdots \to X$$

by composing $\mathcal{BMO}_1(X', \pi^*I, 1, \sum_{i=0}^{k-1} E^i)$ with $X' \to X$. By construction, $\Pi_{(X,I,E)}^* I = I_r \cdot \mathcal{O}_{X_r}(F)$ for some effective divisor F supported on the total transform of $\sum_{i=0}^{k-1} E^i$. Here $I_r = \mathcal{O}_{X_r}$ since max-ord $I_r < 1$ and F is a simple normal crossing divisor. Therefore $\Pi_{(X,I,E)}^* I$ is a monomial ideal which can be written down explicitly as follows.

Let $F_j \subset X_{j+1}$ denote the exceptional divisor of the jth step in the above smooth blow-up sequence for $\Pi_{(X,I,E)} : X_r \to X$. Then

$$\Pi_{(X,I,E)}^* I = \mathcal{O}_{X_r}\big(-\textstyle\sum_{j=s}^{r} \Pi_{r,j+1}^* F_j\big),$$

where $\Pi_{r,j+1} : X_r \to X_{j+1}$ is the corresponding composite of blow-ups.

The functoriality properties required in (3.35) follow from the corresponding functoriality properties in (3.69) and from (3.71.2). $\qquad\square$

3.7. Birational transform of derivatives

DEFINITION 3.73 (Derivative of an ideal sheaf). On a smooth scheme X over a field k, let Der_X denote the sheaf of derivations $\mathcal{O}_X \to \mathcal{O}_X$.

If $x_1, \ldots, x_n$ are local coordinates at a point $p \in X$, then the derivations $\partial/\partial x_1, \ldots, \partial/\partial x_n$ are local generators of Der_X. Derivation gives a k-bilinear map

$$\mathrm{Der}_X \times \mathcal{O}_X \to \mathcal{O}_X.$$

Let $I \subset \mathcal{O}_X$ be an ideal sheaf. Its *first derivative* is the ideal sheaf $D(I)$ generated by all derivatives of elements of I. That is,

$$D(I) := \big(\mathrm{im}[\mathrm{Der}_X \times I \to \mathcal{O}_X]\big). \tag{3.73.1}$$

Note that $I \subset D(I)$, as shown by the formula

$$f = \frac{\partial(xf)}{\partial x} - x\frac{\partial f}{\partial x}.$$

In terms of generators we can write $D(I)$ as

$$D(f_1, \ldots, f_s) = \left(f_i, \frac{\partial f_i}{\partial x_j} : 1 \le i \le s, 1 \le j \le n \right).$$

Higher derivatives are defined inductively by

$$D^{r+1}(I) := D\big(D^r(I)\big). \tag{3.73.2}$$

(Note that $D^r(I)$ contains all rth partial derivatives of elements of I, but over general rings it is bigger; try second derivatives over $\mathbb{Z}[x]$. Over characteristic zero fields, they are actually equal, as one can see using formulas like

$$\frac{\partial f}{\partial y} = \frac{\partial^2 (xf)}{\partial y \partial x} - x \frac{\partial^2 f}{\partial y \partial x} \quad \text{and} \quad 2\frac{\partial f}{\partial x} = \frac{\partial^2 (xf)}{\partial x^2} - x\frac{\partial^2 f}{\partial x^2}.$$

The inductive definition is easier to work with.)

If max-ord$\,I \le m$, then $D^m(I) = \mathcal{O}_X$, and thus the $D^r(I)$ give an ascending chain of ideal sheaves

$$I \subset D(I) \subset D^2(I) \subset \cdots \subset D^m(I) = \mathcal{O}_X.$$

This is, however, not the right way to look at derivatives. Since differentiating a function r times is expected to reduce its order by r, we define the derivative of a marked ideal by

$$D^r(I, m) := \big(D^r(I), m - r\big) \quad \text{for } r \le m. \tag{3.73.3}$$

Before we can usefully compare the ideal I and its higher derivatives, we have to correct for the difference in their markings.

Higher derivatives have the usual properties.

LEMMA 3.74. *Let the notation be as above. Then*

(1) $D^r(D^s(I)) = D^{r+s}(I)$,
(2) $D^r(I \cdot J) \subset \sum_{i=0}^{r} D^i(I) \cdot D^{r-i}(J)$ *(product rule)*,
(3) $\operatorname{cosupp}(I, m) = \operatorname{cosupp}(D^r(I), m - r)$ *for $r < m$ (char. 0 only!)*,
(4) *if $f : Y \to X$ is smooth, then $D(f^*I) = f^*(D(I))$*,
(5) $D(\hat{I}) = \widehat{D(I)}$, *where ^ denotes completion (3.55).* $\qquad\square$

3.74.6 (Aside about positive characteristic). The above definition of higher derivatives is "correct" only in characteristic zero. In general, one should use the *Hasse-Dieudonné derivatives*, which are essentially given by

$$\frac{1}{r_1! \cdots r_n!} \cdot \frac{\partial^{\sum r_i}}{\partial x_1^{r_1} \cdots \partial x_n^{r_n}}.$$

These operators then have other problems. One of the main difficulties of resolution in positive characteristic is a lack of good replacement for higher derivatives.

3.75 (Birational transform of derivatives). Let X be a smooth variety, $Z \subset X$ a smooth subvariety and $\pi : B_Z X \to X$ the blow-up with exceptional divisor $F \subset B_Z X$. Let (I, m) be a marked ideal sheaf on X such that $m \leq \operatorname{ord}_Z I$. Choose local coordinates $(x_1, \ldots, x_n)$ such that $Z = (x_1 = \cdots = x_r = 0)$. Then

$$y_1 = \tfrac{x_1}{x_r}, \ldots, y_{r-1} = \tfrac{x_{r-1}}{x_r}, \ y_r = x_r, \ldots, y_n = x_n$$

are local coordinates on a chart of $B_Z X$. Let us compute the derivatives of

$$\pi_*^{-1}\big(f(x_1, \ldots, x_n), m\big) = \big(y_r^{-m} f(y_1 y_r, \ldots, y_{r-1} y_r, y_r, \ldots, y_n), m\big),$$

defined in (3.60.3). The easy formulas are

$$\begin{aligned}
\tfrac{\partial}{\partial y_j} \pi_*^{-1}(f, m) &= \pi_*^{-1}\big(\tfrac{\partial}{\partial x_j} f, m-1\big) \quad \text{for } j < r, \\
\tfrac{\partial}{\partial y_j} \pi_*^{-1}(f, m) &= \tfrac{1}{y_r} \pi_*^{-1}\big(\tfrac{\partial}{\partial x_j} f, m-1\big) \quad \text{for } j > r,
\end{aligned}$$

and a more complicated one using the chain rule for $j = r$:

$$\begin{aligned}
\tfrac{\partial}{\partial y_r} \pi_*^{-1}(f, m) &= \tfrac{y_i}{y_r} \sum_{i<r} \pi_*^{-1}\big(\tfrac{\partial}{\partial x_i} f, m-1\big) + \tfrac{1}{y_r} \pi_*^{-1}\big(\tfrac{\partial}{\partial x_r} f, m-1\big) \\
&\quad + \big(\tfrac{-m}{y_r}, -1\big) \cdot \pi_*^{-1}(f, m),
\end{aligned}$$

where, as in (3.59), multiplying by $\big(\tfrac{-m}{y_r}, -1\big)$ means multiplying the function by $\tfrac{-m}{y_r}$ and lowering the marking by 1.

These can be rearranged to

$$\pi_*^{-1}\big(\tfrac{\partial}{\partial x_j} f, m-1\big) = \tfrac{\partial}{\partial y_j} \pi_*^{-1}(f, m) \quad \text{for } j < r, \tag{3.75.1}$$

$$\pi_*^{-1}\big(\tfrac{\partial}{\partial x_j} f, m-1\big) = y_r \tfrac{\partial}{\partial y_j} \pi_*^{-1}(f, m) \quad \text{for } j > r, \tag{3.75.2}$$

$$\begin{aligned}
\pi_*^{-1}\big(\tfrac{\partial}{\partial x_r} f, m-1\big) &= y_r \tfrac{\partial}{\partial y_r} \pi_*^{-1}(f, m) - y_r \sum_{i<r} \tfrac{\partial}{\partial y_i} \pi_*^{-1}(f, m) \\
&\quad + (m, -1) \cdot \pi_*^{-1}(f, m).
\end{aligned} \tag{3.75.3}$$

For later purposes, also note the following version of (3.75.1):

$$\pi_*^{-1}\big(x_j \tfrac{\partial}{\partial x_j} f, m-1\big) = y_r y_j \tfrac{\partial}{\partial y_j} \pi_*^{-1}(f, m) \quad \text{for } j < r. \tag{3.75.4}$$

Observe that the right-hand sides of these equations are in $D(\pi_*^{-1}(f, m))$. Thus we have proved the following elementary but important statement.

THEOREM 3.76. *Let (I, m) be a marked ideal and $\Pi : X_r \to X$ the composite of a smooth blow-up sequence of order $\geq m$ starting with (X, I, m). Then*

$$\Pi_*^{-1}\big(D^j(I, m)\big) \subset D^j\big(\Pi_*^{-1}(I, m)\big) \quad \text{for every } j \geq 0.$$

Proof. For $j = 1$ and for one blow-up this is what the above formulas (3.75.1–3) say. The rest follows by induction on j and on the number of blow-ups. $\square$

COROLLARY 3.77. *Let*

$$\Pi : (X_r, I_r, m) \xrightarrow{\pi_{r-1}} (X_{r-1}, I_{r-1}, m) \xrightarrow{\pi_{r-2}} \cdots$$
$$\xrightarrow{\pi_1} (X_1, I_1, m) \xrightarrow{\pi_0} (X_0, I_0, m)$$

be a smooth blow-up sequence of order $\geq m$ starting with (X, I, m).
Fix $j \leq m$, and define inductively the ideal sheaves J_i by

$$J_0 := D^j(I) \quad and \quad (J_{i+1}, m-j) := (\pi_i)_*^{-1}(J_i, m-j).$$

Then, $J_i \subset D^j(I_i)$ for every i, and we get a smooth blow-up sequence of order $\geq m - j$ starting with $(X, D^j(I), m-j)$

$$\Pi : (X_r, J_r, m-j) \xrightarrow{\pi_{r-1}} (X_{r-1}, J_{r-1}, m-j) \xrightarrow{\pi_{r-2}} \cdots$$
$$\xrightarrow{\pi_1} (X_1, J_1, m-j) \xrightarrow{\pi_0} (X_0, J_0, m-j).$$

Proof. We need to check that for every $i < r$ the inequality $\operatorname{ord}_{Z_i} J_i \geq m - j$ holds, where $Z_i \subset X_i$ is the center of the blow-up $\pi_i : X_{i+1} \to X_i$. If $\Pi_i : X_i \to X$ is the composition, then

$$J_i = (\Pi_i)_*^{-1}(D^j I, m-j) \subset D^j\big((\Pi_i)_*^{-1}(I, m)\big) = D^j(I_i, m),$$

where the containment in the middle follows from (3.76). By assumption $\operatorname{ord}_{Z_i} I_i \geq m$, and thus $\operatorname{ord}_{Z_i} D^j(I_i) \geq m - j$ by (3.74.3). $\qquad\square$

3.8. Maximal contact and going down

DEFINITION 3.78. Let X be a smooth variety, $I \subset \mathcal{O}_X$ an ideal sheaf and $m = \text{max-ord}\, I$. A smooth hypersurface $H \subset X$ is called a hypersurface of *maximal contact* if the following holds.

For every open set $X^0 \subset X$ and for every smooth blow-up sequence of order m starting with $(X^0, I^0 := I|_{X^0})$,

$$\Pi : (X_r^0, I_r^0) \xrightarrow{\pi_{r-1}} (X_{r-1}^0, I_{r-1}^0) \xrightarrow{\pi_{r-2}} \cdots \xrightarrow{\pi_1} (X_1^0, I_1^0) \xrightarrow{\pi_0} (X_0^0, I_0^0),$$

the center of every blow-up $Z_i^0 \subset X_i^0$ is contained in the birational transform $H_i^0 \subset X_i^0$ of $H^0 := H \cap X^0$. This implies that

$$\Pi|_{H_r^0} : (H_r^0, I_r|_{H_r^0}, m) \xrightarrow{\pi_{r-1}} (H_{r-1}^0, I_{r-1}|_{H_{r-1}^0}, m) \xrightarrow{\pi_{r-2}} \cdots$$
$$\xrightarrow{\pi_1} (H_1^0, I_1|_{H_1^0}, m) \xrightarrow{\pi_0} (H_0^0, I_0|_{H_0^0}, m)$$

is a smooth blow-up sequence of order $\geq m$ starting with $(H^0, I|_{H^0}, m)$.

Being a hypersurface of maximal contact is a local property.

For now we ignore the divisorial part E of a triple (X, I, E) since we cannot guarantee that $E|_H$ is also a simple normal crossing divisor.

DEFINITION 3.79. Let X be a smooth variety, $I \subset \mathcal{O}_X$ an ideal sheaf and $m = \text{max-ord}\, I$. The *maximal contact ideal* of I is

$$MC(I) := D^{m-1}(I).$$

Note that $MC(I)$ has order 1 at $x \in X$ if $\mathrm{ord}_x I = m$ and order 0 if $\mathrm{ord}_x I < m$. Thus

$$\mathrm{cosupp}\, MC(I) = \mathrm{cosupp}(I, m).$$

THEOREM 3.80 (Maximal contact). *Let X be a smooth variety, $I \subset \mathcal{O}_X$ an ideal sheaf and $m = \max\text{-}\mathrm{ord}\, I$. Let L be a line bundle on X and $h \in H^0(X, L \otimes MC(I))$ a section with zero divisor $H := (h = 0)$.*

(1) *If H is smooth and $I|_H \neq 0$, then H is a hypersurface of maximal contact.*

(2) *Every $x \in X$ has an open neighborhood $x \in U_x \subset X$ and $h_x \in H^0(U_x, L \otimes MC(I))$ such that $H_x := (h_x = 0) \subset U_x$ is smooth.*

Proof. Being a hypersurface of maximal contact is a local question, and thus we may assume that $L = \mathcal{O}_X$. Let

$$\Pi : (X_r, I_r) \xrightarrow{\pi_{r-1}} (X_{r-1}, I_{r-1}) \xrightarrow{\pi_{r-2}} \cdots \xrightarrow{\pi_1} (X_1, I_1) \xrightarrow{\pi_0} (X_0, I_0)$$

be a smooth blow-up sequence of order m starting with (X, I), where π_i is the blow-up of $Z_i \subset X_i$.

Applying (3.77) for $j = m - 1$, we obtain a smooth blow-up sequence of order ≥ 1 starting with $(X, J_0 := MC(I), 1)$:

$$\Pi : (X_r, J_r, 1) \xrightarrow{\pi_{r-1}} (X_{r-1}, J_{r-1}, 1) \xrightarrow{\pi_{r-2}} \cdots \xrightarrow{\pi_1} (X_1, J_1, 1) \xrightarrow{\pi_0} (X_0, J_0, 1).$$

Let $H_i := (\Pi_i)_*^{-1} H \subset X_i$ denote the birational transform of $H \subset X$. Since $\mathcal{O}_{X_0}(-H_0) \subset J_0$ and H_0 is smooth, we see that $\mathcal{O}_{X_i}(-H_i) \subset J_i$ for every i. By assumption $\mathrm{ord}_{Z_i} I_i \geq m$. Thus, using (3.74.3) and (3.77) we get that

$$\mathrm{ord}_{Z_i} J_i \geq \mathrm{ord}_{Z_i} MC(I_i) \geq 1$$

and hence also $\mathrm{ord}_{Z_i} H_i \geq 1$. Thus $Z_i \subset H_i$ for every i, and so H is a hypersurface of maximal contact.

To see the second claim, pick $x \in X$ such that $\mathrm{ord}_x I = m$. Then $\mathrm{ord}_x MC(I) = 1$ by (3.74.3). Thus there is a local section of $MC(I)$ that has order 1 at x, and so its zero divisor is smooth in a neighborhood of x. $\square$

ASIDE 3.81. A section $h \in MC(I)$ such that $H = (h = 0)$ is smooth always exists locally but usually not globally, not even if we tensor I by a very ample line bundle L. By the Bertini-type theorem of [Kol97, 4.4], the best one can achieve globally is that H has cA-type singularities. (These are given by local equations $x_1 x_2 + (\text{other terms}) = 0$.)

The above results say that every smooth blow-up sequence of order m starting with (X, I) can be seen as a smooth blow-up sequence starting with $(H, I|_H, m)$.

An important remaining problem is that not every smooth blow-up sequence starting with $(H, I|_H, m)$ corresponds to a smooth blow-up sequence of order m starting with (X, I), and thus we cannot yet construct an order reduction of (X, I) from an order reduction of $(H, I|_H, m)$.

Here are some examples that show what can go wrong.

EXAMPLE 3.82. Let $I = (xy - z^n)$. Then $\mathrm{ord}_0 I = 2$ and $D(I) = (x, y, z^{n-1})$. $H := (x = 0)$ is a surface of maximal contact, and

$$(H, I|_H) \cong \left(\mathbb{A}^2_{y,z}, (z^n)\right).$$

Thus $(H, I|_H)$ shows a 1-dimensional singular locus of order n, whereas we have an isolated singular point of order 2. The same happens if we use $(y = 0)$ as a surface of maximal contact.

In this case we do better if we use a general surface of maximal contact. Indeed, for $H_g := (x - y = 0)$,

$$(H_g, I|_{H_g}) \cong \left(\mathbb{A}^2_{x,z}, (x^2 - z^n)\right),$$

and we get an equivalence between smooth blow-up sequences of order 2 starting with $(\mathbb{A}^3, (xy - z^n))$ and smooth blow-up sequences of order ≥ 2 starting with $(\mathbb{A}^2, (x^2 - z^n), 2)$.

In some cases, even the general hypersurface of maximal contact fails to produce an equivalence. There are no problems on H itself, but difficulties appear after blow-ups.

Let $I = (x^3 + xy^5 + z^4)$. A general surface of maximal contact is

$$H := (x + u_1 xy^3 + u_2 y^4 + u_3 z^2 = 0), \quad \text{where the } u_i \text{ are units.}$$

Let us compute two blow-ups given by $x_1 = x/y, y_1 = y, z_1 = z/y$ and $x_2 = x_1/y_1, y_2 = y_1, z_2 = z_1/y_1$. We get the birational transforms

$$\begin{array}{ll}
x^3 + xy^5 \ \ + z^4 & x \ + u_1 xy^3 \ + u_2 y^4 + u_3 z^2 \\
x_1^3 + x_1 y_1^3 + y_1 z_1^4 & x_1 + u_1 x_1 y_1^3 + u_2 y_1^3 + u_3 y_1 z_1^2 \\
x_2^3 + x_2 y_2 + y_2^2 z_2^4 & x_2 + u_1 x_2 y_2^3 + u_2 y_2^3 + u_3 y_2^2 z_2^2.
\end{array}$$

The second birational transform of the ideal has order 2 on this chart. However, its restriction to the birational transform H_2 of H still has order 3 since we can use the equation of H_2 to eliminate x_2 by the substitution

$$x_2 = -y_2^2 (u_2 + u_3 z_2^2)(1 + u_1 y_2^3)^{-1}$$

to obtain that $I_2|_{H_2} \subset (y_2^3, y_2^2 z_2^4)$.

3.9. Restriction of derivatives and going up

In general, neither the order of an ideal nor its derivative ideal commute with restrictions to smooth hypersurfaces. For instance, if $I = (x^2 + xy + z^3)$ and $S = (x = 0)$ then $\mathrm{ord}_0 I = 2$ but $\mathrm{ord}_0(I|_S) = 3$ and $(DI)|_S = (y, z^2)$ but $D(I|_S) = (z^2)$. It is easy to see that

$$\mathrm{ord}_p I \leq \mathrm{ord}_p(I|_S) \quad \text{and} \quad (DI)|_S \supset D(I|_S),$$

but neither is an equality. The notion of D-balanced ideals provides a solution to the first of these problems and a partial remedy to the second.

DEFINITION 3.83. As in (3.52), an ideal I with $m = \text{max-ord}\, I$ is called *D-balanced* if

$$\left(D^i I\right)^m \subset I^{m-i} \quad \forall\, i < m.$$

If I is D-balanced, then at every point it has order either m or 0. Indeed, if $\text{ord}_p I < m$ then $(D^{m-1}I)_p = \mathcal{O}_{p,X}$, thus I^{m-1} and I both contain a unit at p. In particular, $\text{cosupp}(I, m) = \text{cosupp}\, I$, hence the maximal order commutes with restrictions.

We can reformulate this observation as follows. If I is D-balanced, then any smooth blow-up of order $\geq m$ for $I|_S$ corresponds to a smooth blow-up of order $\geq m$ for I. We would like a similar statement not just for one blow-up, but for all blow-up sequences.

THEOREM 3.84 (Going-up property of D-balanced ideals). *Let X be a smooth variety and I a D-balanced sheaf of ideals with $m = \text{max-ord}\, I$. Let $S \subset X$ be any smooth hypersurface such that $S \not\subset \text{cosupp}(I, m)$ and*

$$\Pi^S : (S_r, J_r, m) \xrightarrow{\pi^S_{r-1}} (S_{r-1}, J_{r-1}, m) \xrightarrow{\pi^S_{r-2}} \cdots$$
$$\xrightarrow{\pi^S_1} (S_1, J_1, m) \xrightarrow{\pi^S_0} (S_0, J_0, m) = (S, I|_S, m)$$

be a smooth blow-up sequence of order $\geq m$, where π^S_i is the blow-up of $Z_i \subset S_i$. Then the pushed-forward sequence (3.30)

$$\Pi : (X_r, I_r) \xrightarrow{\pi_{r-1}} (X_{r-1}, I_{r-1}) \xrightarrow{\pi_{r-2}} \cdots$$
$$\xrightarrow{\pi_1} (X_1, I_1) \xrightarrow{\pi_0} (X_0, I_0) = (X, I)$$

is a smooth blow-up sequence of order m, where π_i is the blow-up of $Z_i \subset S_i \subset X_i$.

COROLLARY 3.85 (Going up and down). *Let X be a smooth variety, $I \subset \mathcal{O}_X$ a D-balanced ideal sheaf with $m = \text{max-ord}\, I$ and E a divisor with simple normal crossings. Let $H \subset X$ be a smooth hypersurface of maximal contact such that $E + H$ is also a divisor with simple normal crossings and no irreducible component of H is contained in $\text{cosupp}(I, m)$.*

Then pushing forward (3.30) from H to X is a one-to-one correspondence between

(1) *smooth blow-up sequences of order $\geq m$ starting with the triple $(H, I|_H, m, E|_H)$, and*
(2) *smooth blow-up sequences of order m starting with (X, I, E).*

Proof. This follows from (3.84) and (3.80), except for the role played by E.

Adding E to (X, I) (resp., to $(H, m, I|_H)$) means that now we can use only blow-ups whose centers are in simple normal crossing with E (resp.,

$E|_H$) and their total transforms. Since $E|_H$ is again a divisor with simple normal crossings, this poses the same restriction on order reduction for (X, I, E) as on order reduction for $(H, I|_H, m, E|_H)$. $\square$

3.86 (First attempt to prove (3.84)). We already noted that we are ok for the first blow-up. Let us see what happens with pushing forward the second blow-up $\pi_1^S : S_2 \to S_1$. By assumption $\mathrm{ord}_{Z_1}(\pi_0^S)_*^{-1}(I|_S, m) \geq m$. Can we conclude from this that $\mathrm{ord}_{Z_1}(\pi_0)_*^{-1}(I, m) \geq m$? In other words, is

$$S_1 \cap \mathrm{cosupp}\big((\pi_0)_*^{-1}(I, m)\big) = \mathrm{cosupp}\big((\pi_0^S)_*^{-1}(I|_S, m)\big)?$$

Since the birational transform commutes with restrictions, this indeed holds if the birational transform $(\pi_0)_*^{-1}(I, m)$ is again D-balanced. By assumption $(D^i I)^m \subset I^{m-i}$ and so

$$\big((\pi_0)_*^{-1}(D^i I, m - i)\big)^m \subset \big((\pi_0)_*^{-1}(I, m)\big)^{m-i}.$$

Unfortunately, when we interchange $(\pi_0)_*^{-1}$ and D^i on the left-hand side, the inequality in (3.76) goes the wrong way and indeed, in general the birational transform is not D-balanced.

Looking at the formulas (3.75.1–3), we see that taking birational transform commutes with some derivatives but not with others.

In order to exploit this, we introduce logarithmic derivatives. This notion enables us to separate the "good" directions from the "bad" ones.

Example 3.86.1. Check that (x^2, xy^m, y^{m+1}) is D-balanced. After blowing up the origin, one of the charts gives $(x_1^2, x_1 y_1^{m-1}, y_1^{m-1})$, which is not D-balanced.

3.87 (Logarithmic derivatives). Let X be a smooth variety and $S \subset X$ a smooth subvariety. For simplicity, we assume that S is a hypersurface. At a point $p \in S$ pick local coordinates $x_1, \ldots, x_n$ such that $S = (x_1 = 0)$. If f is any function, then

$$\frac{\partial f}{\partial x_i}|_S = \frac{\partial(f|_S)}{\partial x_i} \quad \text{for } i > 1,$$

but $\partial(f|_S)/\partial x_1$ does not even make sense. Therefore, we would like to decompose $D(f)$ into two parts:

- $\partial f/\partial x_i$ for $i > 1$ (these commute with restriction to S), and
- $\partial f/\partial x_1$ (which does not).

Such a decomposition is, however, not coordinate invariant. The best one can do is the following.

Let $\mathrm{Der}_X(-\log S) \subset \mathrm{Der}_X$ be the largest subsheaf that maps $\mathcal{O}_X(-S)$ into itself by derivations. It is called the sheaf of *logarithmic derivations* along S. In the above local coordinates we can write

$$\mathrm{Der}_X(-\log S) = \left(x_1\frac{\partial}{\partial x_1}, \frac{\partial}{\partial x_2}, \ldots, \frac{\partial}{\partial x_n}\right).$$

For an ideal sheaf I set

$$
\begin{aligned}
D(-\log S)(I) &:= \big(\mathrm{im}[\mathrm{Der}_X(-\log S) \times I \to \mathcal{O}_X]\big) \quad \text{and}\\
D^{r+1}(-\log S)(I) &:= D(-\log S)\big(D^r(-\log S)(I)\big) \quad \text{for } r \geq 1.
\end{aligned}
$$

We need three properties of log derivations.

First, log derivations behave well with respect to restriction to S:

$$\big(D^r(-\log S)(I)\big)|_S = D^r(I|_S). \tag{3.87.1}$$

Second, one can filter the sheaf $D^s(I)$ by subsheaves

$$D^s(-\log S)(I) \subset D^{s-1}(-\log S)\big(D(I)\big) \subset \cdots \subset D^s(I).$$

There are no well-defined complements, but in local coordinates $x_1, \ldots, x_n$ we can write

$$D^s(I) = D^s(-\log S)(I) + D^{s-1}(-\log S)\left(\frac{\partial I}{\partial x_1}\right) + \cdots + \left(\frac{\partial^s I}{\partial x_1^s}\right), \tag{3.87.2}$$

and the first $j+1$ summands span $D^{s-j}(-\log S)(D^j(I))$.

Third, under the assumptions of (3.84), we get a logarithmic version of (3.76):

$$\Pi_*^{-1}\Big(D^j(-\log S_r)(I,m)\Big) \subset D^j(-\log S)\Big(\Pi_*^{-1}(I,m)\Big), \tag{3.87.3}$$

which is proved the same way using (3.75.4).

We can now formulate the next result, which can be viewed as a way to reverse the inclusion in (3.76).

THEOREM 3.88. *Consider a smooth blow-up sequence of order $\geq m$:*

$$
\begin{aligned}
\Pi : (X_r, I_r, m) &\xrightarrow{\pi_{r-1}} (X_{r-1}, I_{r-1}, m) \xrightarrow{\pi_{r-2}} \cdots \\
&\xrightarrow{\pi_1} (X_1, I_1, m) \xrightarrow{\pi_0} (X_0, I_0, m) = (X, I, m).
\end{aligned}
$$

Let $S \subset X$ be a smooth hypersurface and $S_i \subset X_i$ its birational transforms. Assume that each blow-up center Z_i is contained in S_i. Then

$$D^s\Pi_*^{-1}(I,m) = \sum_{j=0}^s D^{s-j}(-\log S_r)\Pi_*^{-1}\big(D^j I, m-j\big). \tag{3.88.1}$$

Proof. Using (3.76) we obtain that

$$
\begin{aligned}
D^{s-j}(-\log S_r)\Pi_*^{-1}\big(D^j I, m-j\big) &\subset D^{s-j}\Pi_*^{-1}\big(D^j I, m-j\big)\\
&\subset D^{s-j}D^j(\Pi_*^{-1}I, m) = D^s\Pi_*^{-1}(I, m),
\end{aligned}
$$

and thus the right-hand side of (3.88.1) is contained in the left-hand side.

Next let us check the reverse inclusion in (3.88.1) for one blow-up. The question is local on X, and so choose coordinates $x_1, \ldots, x_n$ such that

$S = (x_1 = 0)$ and the center of the blow-up π is $(x_1 = \cdots = x_r = 0)$. We have a typical local chart

$$y_1 = \tfrac{x_1}{x_r}, \ldots, y_{r-1} = \tfrac{x_{r-1}}{x_r}, \ y_r = x_r, \ldots, y_n = x_n,$$

and $S_1 = (y_1 = 0)$ is the birational transform of S. Note that the blow-up is covered by r different charts, but only $r-1$ of these can be written in the above forms, where x_r is different from x_1. These $r-1$ charts, however, completely cover S_1.

Applying (3.87.2) to $\pi_*^{-1}(I, m)$ we obtain that

$$D^s \pi_*^{-1}(I, m) = \sum_{j=0}^{s} D^{s-j}(-\log S_1)\left(\frac{\partial^j \pi_*^{-1}(I,m)}{\partial y_1^j}\right).$$

Although usually differentiation does not commute with birational transforms, by (3.75.1) it does so for $\partial/\partial x_1$ and $\partial/\partial y_1$. So we can rewrite the above formula as

$$
\begin{aligned}
D^s \pi_*^{-1}(I, m) &= \sum_{j=0}^{s} D^{s-j}(-\log S_1)\pi_*^{-1}\left(\frac{\partial^j (I,m)}{\partial x_1^j}\right) \\
&\subset \sum_{j=0}^{s} D^{s-j}(-\log S_1)\pi_*^{-1}(D^j I, m - j),
\end{aligned}
\tag{3.88.2}
$$

where the inclusion is clear. As noted above, the right-hand side of (3.88.2) is contained in the left-hand side, and hence they are equal. This proves (3.88) for one blow-up.

In the general case, we use induction on the number of blow-ups. We factor $\Pi : X_r \to X$ as the composite of $\pi_{r-1} : X_r \to X_{r-1}$ and $\Pi_{r-1} : X_{r-1} \to X$. Use (3.88) for π_{r-1} to get that

$$
\begin{aligned}
D^s \Pi_*^{-1}(I, m) &= D^s (\pi_{r-1})_*^{-1}(\Pi_{r-1})_*^{-1}(I, m) \\
&= \sum_{j=0}^{s} D^{s-j}(-\log S_r)(\pi_{r-1})_*^{-1} D^j (\Pi_{r-1})_*^{-1}(I, m).
\end{aligned}
\tag{3.88.3}
$$

By induction (3.88) holds for Π_{r-1} and $s = j$, thus

$$D^j (\Pi_{r-1})_*^{-1}(I, m) = \sum_{\ell=0}^{j} D^{j-\ell}(-\log S_{r-1})(\Pi_{r-1})_*^{-1}(D^\ell I, m - \ell).$$

By (3.87.3), we can interchange $(\pi_{r-1})_*^{-1}$ and $D^{j-\ell}(-\log S_{r-1})$, and so

$$
\begin{aligned}
(\pi_{r-1})_*^{-1} D^j &(\Pi_{r-1})_*^{-1}(I, m) \\
&= (\pi_{r-1})_*^{-1} \sum_{\ell=0}^{j} D^{j-\ell}(-\log S_{r-1})(\Pi_{r-1})_*^{-1}(D^\ell I, m - \ell) \\
&\subset \sum_{\ell=0}^{j} D^{j-\ell}(-\log S_r)(\pi_{r-1})_*^{-1}(\Pi_{r-1})_*^{-1}(D^\ell I, m - \ell) \\
&= \sum_{\ell=0}^{j} D^{j-\ell}(-\log S_r)\Pi_*^{-1}(D^\ell I, m - \ell).
\end{aligned}
$$

Substituting into (3.88.3), we obtain the desired result:

$$D^s \Pi_*^{-1}(I, m)$$
$$\subset \sum_{j=0}^{s} D^{s-j}(-\log S_r) \sum_{\ell=0}^{j} D^{j-\ell}(-\log S_r) \Pi_*^{-1}(D^\ell I, m - \ell)$$
$$= \sum_{\ell=0}^{s} D^{s-\ell}(-\log S_r) \Pi_*^{-1}(D^\ell I, m - \ell). \quad \square$$

COROLLARY 3.89. *Let the notation and assumptions be as in (3.88).*
Then

$$S_r \cap \operatorname{cosupp}\big(\Pi_*^{-1}(I, m)\big) = \bigcap_{j=0}^{m-1} \operatorname{cosupp}(\Pi|_{S_r})_*^{-1}\big((D^j I)|_S, m - j\big).$$

Proof. Restricting (3.88.1) to S_r and using (3.62) and (3.87.1) we get
that

$$\big(D^s \Pi_*^{-1}(I, m)\big)|_{S_r} = \sum_{j=0}^{s} D^{s-j}(\Pi|_{S_r})_*^{-1}\big((D^j I)|_S, m - j\big). \qquad (3.89.1)$$

Set $s = m - 1$ and take cosupports. Since $D^{m-1}\Pi_*^{-1}(I, m)$ has order 1,
its cosupport commutes with restrictions, so the left-hand side of (3.89.1)
becomes

$$\begin{aligned}
\operatorname{cosupp}\Big(\big(D^{m-1}\Pi_*^{-1}(I, m)\big)|_{S_r}\Big) &= \operatorname{cosupp}\big(D^{m-1}\Pi_*^{-1}(I, m)\big) \cap S_r \\
&= \operatorname{cosupp}\big(\Pi_*^{-1}(I, m)\big) \cap S_r,
\end{aligned}$$
$$(3.89.2)$$

where the second equality follows from (3.74.3).

On the right-hand side of (3.89.1) use (3.59.4) to obtain that

$$\begin{aligned}
\operatorname{cosupp}\Big(\sum_{j=0}^{m-1} D^{m-1-j}(\Pi|_{S_r})_*^{-1}\big((D^j I)|_S, m - j\big)\Big) & \\
= \bigcap_{j=0}^{m-1} \operatorname{cosupp}\Big(D^{m-1-j}(\Pi|_{S_r})_*^{-1}\big((D^j I)|_S, m - j\big)\Big) & \quad (3.89.3) \\
= \bigcap_{j=0}^{m-1} \operatorname{cosupp}(\Pi|_{S_r})_*^{-1}\big((D^j I)|_S, m - j\big). &
\end{aligned}$$

The last lines of (3.89.2) and (3.89.3) are thus equal. $\qquad \square$

3.90 (Proof of (3.84)). By induction, assume that this already holds for
blow-up sequences of length $< r$. We need to show that the last blow-up
also has order $\geq m$, or, equivalently, $\operatorname{cosupp}(J_{r-1}, m) \subset \operatorname{cosupp}(I_{r-1}, m)$.

Using first (3.89) for $\Pi_{r-1} : X_{r-1} \to X$, then the D-balanced property
in line 2, we obtain that

$$\begin{aligned}
S_{r-1} \cap \operatorname{cosupp}(I_{r-1}, m) &= \bigcap_{j=0}^{m-1} \operatorname{cosupp}(\Pi_{r-1}^S)_*^{-1}\big((D^j I)|_S, m - j\big) \\
&= \bigcap_{j=0}^{m-1} \operatorname{cosupp}(\Pi_{r-1}^S)_*^{-1}\big((D^j I)^m|_S, m(m - j)\big) \\
&\supset \bigcap_{j=0}^{m-1} \operatorname{cosupp}(\Pi_{r-1}^S)_*^{-1}\big(I^{m-j}|_S, m(m - j)\big) \\
&= \bigcap_{j=0}^{m-1} \operatorname{cosupp}(\Pi_{r-1}^S)_*^{-1}\big(I|_S, m\big) \\
&= \operatorname{cosupp}(\Pi_{r-1}^S)_*^{-1}\big(J_0, m\big) \\
&= \operatorname{cosupp}(J_{r-1}, m). \quad \square
\end{aligned}$$

3.10. Uniqueness of maximal contact

Given (X, I, E), let $j : H \hookrightarrow X$ and $j' : H' \hookrightarrow X$ be two hypersurfaces of maximal contact. By (3.85) we can construct smooth blow-up sequences for (X, I, E) from $(H, I|_H, m, E_H)$ and also from $(H', I|_{H'}, m, E_{H'})$. We need to guarantee that we get the same blow-up sequences.

Assume that there is an automorphism ϕ of X such that $\phi^* I = I$ and $\phi^{-1}(E + H') = E + H$. Then $(H, I_H, m, E_H) = \phi^*(H', I|_{H'}, m, E_{H'})$, thus if $\mathcal{B}(H', I|_{H'}, m, E_{H'})$ is the smooth blow-up sequence constructed using H', then the "same" construction using H gives

$$\mathcal{B}(H, I_H, m, E_H) = \phi^* \, \mathcal{B}(H', I|_{H'}, m, E_{H'}).$$

Pushing these forward as in (3.85), we obtain that

$$j_* \mathcal{B}(H, I_H, m, E_H) = \phi^* \big(j'_* \mathcal{B}(H', I|_{H'}, m, E_{H'}) \big).$$

That is, the blow-up sequences we get from H and H' are isomorphic, but we would like them to be identical.

Let Z_0 (resp., Z_0') be the center of the first blow-up obtained using H (resp., H'). As above, $\phi^{-1}(Z_0') = Z_0$. Both Z_0' and Z_0 are contained in $\mathrm{cosupp}(I, m)$, so if ϕ is the identity on $\mathrm{cosupp}(I, m)$ then $Z_0' = Z_0$.

The assumption $\phi^* I = I$ implies that ϕ maps $\mathrm{cosupp}(I, m)$ into itself, but it does not imply that ϕ is the identity on $\mathrm{cosupp}(I, m)$. How can we achieve the latter? Let R be a ring, $J \subset R$ an ideal and σ an automorphism of R. It is easy to see that $\sigma(J) = J$ and σ induces the identity automorphism on R/J iff $r - \sigma(r) \in J$ for every $r \in R$.

How should we choose the ideal J in our situation? It turns out that $J = I$ does not work and the ideal sheaf of $\mathrm{cosupp}(I, m)$ behaves badly for blow-ups. An intermediate choice is given by $D^{m-1}(I) = MC(I)$, which works well.

Another twist is that usually X itself has no automorphisms (not even Zariski locally), so we have to work in a formal or étale neighborhood of a point $x \in X$. (See (3.55) for completions.)

DEFINITION 3.91. Let X be a smooth variety, $p \in X$ a point, I an ideal sheaf such that $\max\text{-}\mathrm{ord}\, I = \mathrm{ord}_p I = m$ and $E = E^1 + \cdots + E^s$ a simple normal crossing divisor. Let $H, H' \subset X$ be two hypersurfaces of maximal contact.

We say that H and H' are *formally equivalent at* p with respect to (X, I, E) if there is an automorphism $\phi : \hat{X} \to \hat{X}$ which moves $(X, I, H + E)$ into $(X, I, H' + E)$ and ϕ is close to the identity. That is,

(1) $\phi(\hat{H}) = \hat{H}'$,
(2) $\phi^*(\hat{I}) = \hat{I}$,
(3) $\phi(\hat{E}^i) = \hat{E}^i$ for $i = 1, \ldots, s$, and
(4) $h - \phi^*(h) \in MC(\hat{I})$ for every $h \in \hat{\mathcal{O}}_{x, X}$.

While this is the important concept, it is somewhat inconvenient to use since we defined resolution, order reduction, and so on for schemes of finite type and not for general schemes like $\hat{X}$.

Even very simple formal automorphisms cannot be realized as algebraic automorphisms on some étale cover. (Check this for the map $x \mapsto \sqrt{x}$, which is a formal automorphism of $(1 \in \hat{\mathbb{C}})$.) Thus we need a slightly modified definition.

We say that H and H' are *étale equivalent* with respect to (X, I, E) if there are étale surjections $\psi, \psi' : U \rightrightarrows X$ such that

  (1') $\psi^{-1}(H) = \psi'^{-1}(H')$,
  (2') $\psi^*(I) = \psi'^*(I)$,
  (3') $\psi^{-1}(E^i) = \psi'^{-1}(E^i)$ for $i = 1, \ldots, s$, and
  (4') $\psi^*(h) - \psi'^*(h) \in MC\big(\psi^*(I)\big)$ for every $h \in \mathcal{O}_X$.

The connection with the formal case comes from noting that ψ is invertible after completion, and then $\phi := \hat{\psi}' \circ \hat{\psi}^{-1} : \hat{X} \to \hat{X}$ is the automorphism we seek.

A key observation of [Wło05] is that for certain ideals I any two smooth hypersurfaces of maximal contact are formal and étale equivalent. Recall (3.53) that an ideal I is MC-invariant if

$$MC(I) \cdot D(I) \subset I,$$

where $MC(I)$ is the ideal of maximal contacts defined in (3.51.2). Since taking derivatives commutes with completion (3.74.5), we see that $\widehat{MC(I)} = MC(\hat{I})$.

THEOREM 3.92 (Uniqueness of maximal contact). *Let X be a smooth variety over a field of characteristic zero, I an MC-invariant ideal sheaf, $m = \max\text{-ord}\, I$ and E a simple normal crossing divisor. Let $H, H' \subset X$ be two smooth hypersurfaces of maximal contact for I such that $H + E$ and $H' + E$ both have simple normal crossings.*

Then H and H' are étale equivalent with respect to (X, I, E).

We start with a general result relating automorphisms and derivations of complete local rings. Since derivations are essentially the first order automorphisms, it is reasonable to expect that an ideal is invariant under a subgroup of automorphisms iff it is invariant to first order. We are, however, in an infinite-dimensional setting, so it is safer to work out the details.

NOTATION 3.93. Let k be a field of characteristic zero, K/k a finite field extension and $R = K[[x_1, \ldots, x_n]]$ the formal power series ring in n variables with maximal ideal m, viewed as a k-algebra. For $g_1, \ldots, g_n \in m$ the map $g : x_i \mapsto g_i$ extends to an automorphism of $R \Leftrightarrow g : m/m^2 \to m/m^2$ is an isomorphism $\Leftrightarrow$ the linear parts of the g_i are linearly independent.

Let $B \subset m$ be an ideal. For $b_i \in B$ the map $g : x_i \mapsto x_i + b_i$ need not generate an automorphism, but $g : x_i \mapsto x_i + \lambda_i b_i$ gives an automorphism for general $\lambda_i \in k$. We call these *automorphisms of the form* $\mathbf{1} + B$.

PROPOSITION 3.94. *Let the notation be as above, and let* $I \subset R$ *be an ideal. The following are equivalent:*

(1) I *is invariant under every automorphism of the form* $\mathbf{1} + B$,
(2) $B \cdot D(I) \subset I$,
(3) $B^j \cdot D^j(I) \subset I$ *for every* $j \geq 1$.

Proof. Assume that $B^j \cdot D^j(I) \subset I$ for every $j \geq 1$. Given any $f \in I$, we need to prove that $f(x_1 + b_1, \ldots, x_n + b_n) \in I$. Take the Taylor expansion

$$f(x_1 + b_1, \ldots, x_n + b_n) = f(x_1, \ldots, x_n) + \sum_i b_i \frac{\partial f}{\partial x_i} + \frac{1}{2} \sum_{i,j} b_i b_j \frac{\partial^2 f}{\partial x_i \partial x_j} + \cdots.$$

For any $s \geq 1$, this gives that

$$f(x_1 + b_1, \ldots, x_n + b_n) \in I + B \cdot D(I) + \cdots + B^s \cdot D^s(I) + m^{s+1} \subset I + m^{s+1}$$

since $B^j \cdot D^j(I) \subset I$ by assumption. Letting s go to infinity, by Krull's intersection theorem (3.55) we conclude that $f(x_1 + b_1, \ldots, x_n + b_n) \in I$.

Conversely, for any $b \in B$ and general $\lambda_i \in k$, invariance under the automorphism $(x_1, x_2, \ldots, x_n) \mapsto (x_1 + \lambda_i b, x_2, \ldots, x_n)$ gives that

$$f(x_1 + \lambda_i b, x_2, \ldots, x_n) = f(x_1, \ldots, x_n) + \lambda_i b \frac{\partial f}{\partial x_1} + \cdots (\lambda_i b)^s \frac{\partial f^s}{\partial x_1^s} \in I + m^{s+1}.$$

Use s different values $\lambda_1, \ldots, \lambda_s$. Since the Vandermonde determinant (λ_i^j) is invertible, we conclude that

$$b \frac{\partial f}{\partial x_1} \in I + m^{s+1}.$$

Letting s go to infinity, we obtain that $B \cdot D(I) \subset I$.

Finally, we prove by induction that $B^j \cdot D^j(I) \subset I$ for every $j \geq 1$. $B^{j+1} \cdot D^{j+1}(I)$ is generated by elements of the form $b_0 \cdots b_j \cdot D(g)$, where $g \in D^j(I)$. The product rule gives that

$$\begin{aligned}
b_0 \cdots b_j \cdot D(g) &= b_0 \cdot D(b_1 \cdots b_j \cdot g) - \sum_{i \geq 1} D(b_i) \cdot (b_0 \cdots \widehat{b_i} \cdots b_j \cdot g) \\
&\in B \cdot D\big(B^j \cdot D^j(I)\big) + B^j \cdot D^j(I) \\
&\subset B \cdot D(I) + B^j \cdot D^j(I) \subset I,
\end{aligned}$$

where the entry $\widehat{b_i}$ is omitted from the products.   □

3.95 (Proof of (3.92)). Let us start with formal equivalence.

Pick local sections $x_1, x_1' \in MC(I)$ such that $H = (x_1 = 0)$ and $H' = (x_1' = 0)$. Choose other local coordinates $x_2, \ldots, x_{s+1}$ at p such that $E^i = (x_{i+1} = 0)$ for $i = 1, \ldots, s$. For a general choice of $x_{s+2}, \ldots, x_n$, we see that $x_1, x_2, \ldots, x_n$ and $x_1', x_2, \ldots, x_n$ are both local coordinate systems.

If X is a k-variety and the residue field of $p \in X$ is K, then $\widehat{\mathcal{O}}_{p,X} \cong K[[x_1, \ldots, x_n]]$ by (3.55), so the computations of (3.94) apply.

Since $x_1 - x_1' \in MC(I)$, the automorphism

$$\phi^*(x_1', x_2, \ldots, x_n) = \left(x_1' + (x_1 - x_1'), x_2, \ldots, x_n\right) = (x_1, x_2, \ldots, x_n)$$

is of the form $\mathbf{1} + MC(I)$. Hence by (3.94) we conclude that $\phi^*(\hat{I}) = \hat{I}$. By construction $\phi(\hat{H}) = \hat{H}'$, $\phi(\hat{E}^i) = \hat{E}^i$ and (3.91.4) is also clear.

In order to go from the formal to the étale case, the key point is to realize the automorphism ϕ on some étale neighborhood. Existence follows from the general approximation theorems of [Art69], but in our case the choice is clear.

Take $X \times X$, and for some $p \in X$ let $x_{11}, x_{12}, \ldots, x_{1n}$ be the corresponding local coordinates on the first factor and $x_{21}', x_{22}, \ldots, x_{2n}$ on the second factor. Set

$$U_1(p) := (x_{11} - x_{21}' = x_{12} - x_{22} = \cdots = x_{1n} - x_{2n} = 0) \subset X \times X.$$

The completion of $U_1(p)$ at (p, p) is the graph of ϕ_p. By shrinking $U_1(p)$, we get $(p, p) \in U_2(p) \subset U_1(p)$ such that both coordinate projections $\psi_p, \psi_p' : U_2(p) \rightrightarrows X$ are étale.

From our previous considerations, we know that (3.91.1–4) hold after taking completions at (p, p). Thus (3.91.1'–4') also hold in an open neighborhood $U(p) \ni (p, p)$ by (3.55).

The images of finitely many of the $U(p)$ cover X. We can take U to be their disjoint union.   □

In Section 3.12 we use the maximal contact hypersurfaces H, H' to construct blow-up sequences $\mathbf{B}$ and $\mathbf{B}'$ which become isomorphic after pulling back to U. The next result shows that they are the same already on X. That is, our blow-ups do not depend on the choice of a hypersurface of maximal contact.

DEFINITION 3.96. Let X be a smooth variety over a field of characteristic zero and let

$$\mathbf{B} \quad := \quad (X_r, I_r) \xrightarrow{\pi_{r-1}} \cdots \xrightarrow{\pi_0} (X_0, I_0) = (X, I), \quad \text{and}$$

$$\mathbf{B}' \quad := \quad (X_r', I_r') \xrightarrow{\pi_{r-1}'} \cdots \xrightarrow{\pi_0'} (X_0', I_0') = (X, I)$$

be two blow-up sequences of order $m = \text{max-ord}\, I$. We say that $\mathbf{B}$ and $\mathbf{B}'$ are *étale equivalent* if there are étale surjections $\psi, \psi' : U \rightrightarrows X$ such that

(1) $\psi^*(I) = \psi'^*(I)$,
(2) $\psi^*(h) - \psi'^*(h) \in MC\left(\psi^*(I)\right)$ for every $h \in \mathcal{O}_X$, and
(3) $\psi^* \mathbf{B} = \psi'^* \mathbf{B}'$.

THEOREM 3.97. *Let X be a smooth variety over a field of characteristic zero and I an MC-invariant ideal sheaf. Let $\mathbf{B}$ and $\mathbf{B}'$ be two blow-up sequences of order $m = \text{max-ord}\, I$ which are étale equivalent.*
    Then $\mathbf{B} = \mathbf{B}'$.

Proof. By assumption there are étale surjections $\psi, \psi' : U \rightrightarrows X$ such that $\psi^* \mathbf{B} = \psi'^* \mathbf{B}'$. Let

$$\mathbf{B}^U := (U_r, I_r^U) \xrightarrow{\pi_{r-1}^U} \cdots \xrightarrow{\pi_0^U} (U_0, I_0^U) = (U, \psi^* I = \psi'^* I)$$

be the common pullback. We prove by induction on i the following claims.

(1) $(X_i, I_i) = (X_i', I_i')$.
(2) ψ, ψ' lift to étale surjections $\psi_i, \psi'_i : U_i \rightrightarrows X_i$ such that

$$\text{im}(\psi_i^* - \psi_i'^*) \subset (\Pi_i^U)_*^{-1}\big(MC(I_0^U), 1\big).$$

(3) $Z_{i-1} = Z_{i-1}'$.

For $i = 0$ there is nothing to prove. Let us see how to go from i to $i+1$. Set $W_i = \text{cosupp}(\Pi_i^U)_*^{-1}\big(MC(I_0^U), 1\big)$ and note that $Z_i^U \subset W_i$ by (3.77). By the inductive assumption (2) $\psi_i|_{W_i} = \psi_i'|_{W_i}$, thus $Z_i = \psi_i(Z_i^U) = \psi_i'(Z_i^U) = Z_i'$. This in turn implies that $X_{i+1} = X_{i+1}'$.

In order to compute the lifting of ψ_i and ψ_i', pick local coordinates $x_1, \ldots, x_n$ on $X_i = X_i'$ such that $Z_i = Z_i' = (x_1 = x_2 = \cdots = x_k = 0)$. By induction,

$$\psi_i'^*(x_j) = \psi_i^*(x_j) - b(i,j) \quad \text{for some } b(i,j) \in (\Pi_i^U)_*^{-1}\big(MC(I^U), 1\big).$$

On the blow-up $\pi_i : X_{i+1} \to X_i$ consider the local chart

$$y_1 = \frac{x_1}{x_r}, \ldots, y_{r-1} = \frac{x_{r-1}}{x_r}, \ y_r = x_r, \ldots, y_n = x_n.$$

We need to prove that

$$\psi_{i+1}^*(y_j) - \psi_{i+1}'^*(y_j) \in (\Pi_{i+1}^U)_*^{-1}\big(MC(I^U), 1\big)$$

for every j. This is clear if $y_j = x_j$, that is, for $j \geq r$. Next we compute the case when $j < r$.

The $b(i,j)$ vanish along Z_i^U and so $(\pi_i^U)^* b(i,j) = \psi_{i+1}^*(x_r) b(i+1,j)$ for some $b(i+1,j) \in (\Pi_{i+1}^U)_*^{-1}\big(MC(I^U), 1\big)$. Hence, for $j < r$, we obtain that

$$
\begin{aligned}
\psi_{i+1}'^*(y_j) &= (\pi_i^U)^* \frac{\psi_i'^*(x_j)}{\psi_i'^*(x_r)} = (\pi_i^U)^* \frac{\psi_i^*(x_j) - b(i,j)}{\psi_i^*(x_r) - b(i,r)} \\
&= \frac{\psi_{i+1}^*(x_j) - \psi_{i+1}^*(x_r) b(i+1,j)}{\psi_{i+1}^*(x_r) - \psi_{i+1}^*(x_r) b(i+1,r)} \\
&= \frac{\psi_{i+1}^*(y_j) - b(i+1,j)}{1 - b(i+1,r)}.
\end{aligned}
$$

This implies that

$$\psi_{i+1}^*(y_j) - {\psi'_{i+1}}^*(y_j) = \frac{b(i+1,j) - b(i+1,r)\psi_{i+1}^*(y_j)}{1 - b(i+1,r)}$$

is in $(\Pi_{i+1}^U)_*^{-1}\big(MC(I_0^U),1\big)$, as required.   □

3.11. Tuning of ideals

Following (3.14.9) and (3.77), we are looking for ideals that contain information about all derivatives of I with equalized markings.

DEFINITION 3.98 (Maximal coefficient ideals). Let X be a smooth variety, $I \subset \mathcal{O}_X$ an ideal sheaf and $m = \text{max-ord}\, I$. The *maximal coefficient ideal* of order s of I is

$$W_s(I) := \left(\prod_{j=0}^{m} \big(D^j(I)\big)^{c_j} : \sum (m-j)c_j \geq s \right) \subset \mathcal{O}_X.$$

The ideals $W_s(I)$ satisfy a series of useful properties.

PROPOSITION 3.99. *Let X be a smooth variety, $I \subset \mathcal{O}_X$ an ideal sheaf and $m = \text{max-ord}\, I$. Then*

(1) $W_{s+1}(I) \subset W_s(I)$ *for every s,*
(2) $W_s(I) \cdot W_t(I) \subset W_{s+t}(I)$,
(3) $D(W_{s+1}(I)) = W_s(I)$,
(4) $MC(W_s(I)) = W_1(I) = MC(I)$,
(5) $W_s(I)$ *is MC-invariant,*
(6) $W_s(I) \cdot W_t(I) = W_{s+t}(I)$ *whenever $t \geq (m-1) \cdot \text{lcm}(2,\ldots,m)$ and $s = r \cdot \text{lcm}(2,\ldots,m)$,*
(7) $\big(W_s(I)\big)^j = W_{js}(I)$ *whenever $s = r \cdot \text{lcm}(2,\ldots,m)$ for some $r \geq m-1$, and*
(8) $W_s(I)$ *is D-balanced whenever $s = r \cdot \text{lcm}(2,\ldots,m)$ for some $r \geq m-1$.*

Proof. Assertions (1) and (2) are clear and the inclusion $D(W_{s+1}(I)) \subset W_s(I)$ follows from the product rule. Pick $x_1 \in MC(I)$ that has order 1 at p. Then $x_1^{s+1} \in W_{s+1}(I)$, implying

$$x_1^s = (s+1)^{-1} \tfrac{\partial}{\partial x_1} x_1^{s+1} \in D\big(W_{s+1}(I)\big).$$

Next we prove by induction on t that $x_1^{s-t} W_t \subset D\big(W_{s+1}(I)\big)$, which gives (3) for $t = s$.

Note that $x_1^{s+1-t} f \in W_{s+1}(I)$ for any $f \in W_t(I)$. Thus

$$(s+1-t)x_1^{s-t} f + x_1^{s+1-t}\big(\tfrac{\partial}{\partial x_1} f\big) = \tfrac{\partial}{\partial x_1}\big(x_1^{s+1-t} f\big) \in D\big(W_{s+1}(I)\big).$$

Since $\tfrac{\partial}{\partial x_1} f \in W_{t-1}(I)$, then by induction, $x_1^{s+1-t}\big(\tfrac{\partial}{\partial x_1} f\big) \in D\big(W_{s+1}(I)\big)$. Hence also $x_1^{s-t} f \in D\big(W_{s+1}(I)\big)$.

Applying (3) repeatedly gives that $MC(W_s(I)) = W_1(I)$, which in turn contains $D^{m-1}(I) = MC(I)$ by definition. Conversely, $W_1(I)$ is generated by products of derivatives, at least one of which is a derivative of order $< m$. Thus

$$W_1(I) \subset \sum_{j<m} D^j(I) = D^{m-1}(I),$$

proving (4). Together with (2) and (3), this implies (5).

Thinking of elements of $D^{m-j}(I)$ as variables of degree j, (6) is implied by (3.99.9), and (7) is a special case of (6).

Finally, if $s = r \cdot \mathrm{lcm}(2, \ldots, m)$ for some $r \geq m - 1$, then using (3) and (7) we get that

$$\left(D^i \left(W_s(I) \right) \right)^s = \left(W_{s-i}(I) \right)^s \subset W_{s(s-i)}(I) = \left(W_s(I) \right)^{s-i}. \quad \square$$

Claim 3.99.9. Let $u_1, \ldots, u_m$ be variables such that $\deg(u_i) = i$. Then any monomial $U = \prod u_i^{c_i}$ with $\deg(U) \geq (r + m - 1) \cdot \mathrm{lcm}(2, \ldots, m)$ can be written as $U = U_1 \cdot U_2$, where $\deg(U_1) = r \cdot \mathrm{lcm}(2, \ldots, m)$.

Proof. Set $V_i = u_i^{\mathrm{lcm}(2,\ldots,m)/i}$, and write $u_i^{c_i} = V_i^{b_i} \cdot W_i$ for some b_i such that $\deg W_i < \mathrm{lcm}(2, \ldots, m)$.

If $\sum b_i \geq r$, then choose $0 \leq d_i \leq b_i$ such that $\sum d_i = r$, and take $U_1 = \prod V_i^{b_i}$. Otherwise, $\deg U < (r-1) \cdot \mathrm{lcm}(2, \ldots, m) + m \cdot \mathrm{lcm}(2, \ldots, m)$, a contradiction. $\square$

Aside 3.99.10. Note that one can think of (3.99.9) as a statement about certain multiplication maps

$$H^0(X, \mathcal{O}_X(a)) \times H^0(X, \mathcal{O}_X(b)) \to H^0(X, \mathcal{O}_X(a+b)),$$

where X is the weighted projective space $\mathbb{P}(1, 2, \ldots, m)$. The above claim is a combinatorial version of the Castelnuovo-Mumford regularity theorem in this case (cf. [Laz04, Sec.1.8]).

It seems to me that (3.99.6) should hold for $t \geq \mathrm{lcm}(2, \ldots, m)$ and even for many smaller values of t as well.

It is easy to see that $(m - 1) \cdot \mathrm{lcm}(2, \ldots, m) \leq m!$ for $m \geq 6$, and one can check by hand that (3.99.6) holds for $t \geq m!$ for $m = 1, 2, 3, 4, 5$. Thus we conclude that $W_{m!}(I)$ is D-balanced. This is not important, but the traditional choice of the coefficient ideal corresponds to $W_{m!}(I)$.

The following close analog of (3.77) leads to ideal sheaves that behave the "same" as a given ideal I, as far as order reduction is concerned.

THEOREM 3.100 (Tuning of ideals, I). *Let X be a smooth variety, $I \subset \mathcal{O}_X$ an ideal sheaf and $m = \mathrm{max\text{-}ord}\, I$. Let $s \geq 1$ be an integer and J any ideal sheaf satisfying*

$$I^s \subset J \subset W_{ms}(I).$$

Then a smooth blow-up sequence

$$X_r \xrightarrow{\pi_{r-1}} X_{r-1} \xrightarrow{\pi_{r-2}} \cdots \xrightarrow{\pi_1} X_1 \xrightarrow{\pi_0} X_0 = X$$

is a smooth blow-up sequence of order $\geq m$ starting with (X, I, m) iff it is a smooth blow-up sequence of order $\geq ms$ starting with (X, J, ms).

Proof. Assume that we get a smooth blow-up sequence starting with (X, I, m):

$$(X_r, I_r, m) \xrightarrow{\pi_{r-1}} (X_{r-1}, I_{r-1}, m) \xrightarrow{\pi_{r-2}} \cdots$$
$$\xrightarrow{\pi_1} (X_1, I_1, m) \xrightarrow{\pi_0} (X_0, I_0, m) = (X, I, m).$$

We prove by induction on r that we also get a smooth blow-up sequence starting with (X, J, ms):

$$(X_r, J_r, ms) \xrightarrow{\pi_{r-1}} (X_{r-1}, J_{r-1}, ms) \xrightarrow{\pi_{r-2}} \cdots$$
$$\xrightarrow{\pi_1} (X_1, J_1, ms) \xrightarrow{\pi_0} (X_0, J_0, ms) = (X, J, ms).$$

Assume that this holds up to step $r - 1$. We need to show that the last blow-up $\pi_{r-1} : X_r \to X_{r-1}$ is a blow-up for (X_{r-1}, J_{r-1}, ms). That is, we need to show that

$$\mathrm{ord}_Z \, I_{r-1} \geq m \;\Rightarrow\; \mathrm{ord}_Z \, J_{r-1} \geq ms \quad \text{for any } Z \subset X_{r-1}.$$

Let $\Pi_{r-1} : X_{r-1} \to X_0$ denote the composite. Since $J \subset W_{ms}(I)$, we know that

$$
\begin{aligned}
J_{r-1} &= \left(\Pi_{r-1}\right)_*^{-1}(J, ms) \\
&\subset \left(\Pi_{r-1}\right)_*^{-1}\left(W_{ms}(I), ms\right) \\
&= \left(\Pi_{r-1}\right)_*^{-1}\left(\prod_j \left(D^j I, m - j\right)^{c_j} : \textstyle\sum (m - j)c_j \geq ms\right) \\
&= \left(\prod_j \left((\Pi_{r-1})_*^{-1}(D^j I, m - j)\right)^{c_j} : \textstyle\sum (m - j)c_j \geq ms\right) \\
&\subset \left(\prod_j \left(D^j (\Pi_{r-1})_*^{-1}(I, m)\right)^{c_j} : \textstyle\sum (m - j)c_j \geq ms\right) \quad \text{by (3.76)} \\
&= \left(\prod_j \left(D^j (I_{r-1}, m)\right)^{c_j} : \textstyle\sum (m - j)c_j \geq ms\right).
\end{aligned}
$$

If $\mathrm{ord}_Z \, I_{r-1} \geq m$, then $\mathrm{ord}_Z \, D^j(I_{r-1}) \geq m - j$, and so

$$\mathrm{ord}_Z \prod_j \left(D^j(I_{r-1}, m)\right)^{c_j} \geq \sum (m - j)c_j \geq ms,$$

proving one direction.

In order to prove the converse, let

$$(X_r, J_r, ms) \xrightarrow{\pi_{r-1}} (X_{r-1}, J_{r-1}, ms) \xrightarrow{\pi_{r-2}} \cdots$$
$$\xrightarrow{\pi_1} (X_1, J_1, ms) \xrightarrow{\pi_0} (X_0, J_0, ms) = (X, J, ms)$$

be a smooth blow-up sequence starting with (X, J, ms). Again by induction we show that it gives a smooth blow-up sequence starting with (X, I, m).

Since $I^s \subset J$, we know that

$$I^s_{r-1} = \left((\Pi_{r-1})^{-1}_* I\right)^s \subset (\Pi_{r-1})^{-1}_* (J, ms) = J_{r-1}.$$

Thus if $\mathrm{ord}_Z J_{r-1} \geq ms$, then $\mathrm{ord}_Z I_{r-1} \geq m$, and so $\pi_{r-1} : X_r \to X_{r-1}$ is also a blow-up for (X_{r-1}, I_{r-1}, m). $\square$

COROLLARY 3.101 (Tuning of ideals, II). *Let X be a smooth variety, $I \subset \mathcal{O}_X$ an ideal sheaf with $m = \max\text{-}\mathrm{ord}\, I$ and E a divisor with simple normal crossings. Let $s = r \cdot \mathrm{lcm}(2, \ldots, m)$ for some $r \geq m - 1$. Then $W_s(I)$ is MC-invariant, D-balanced, and a smooth blow-up sequence*

$$X_r \xrightarrow{\pi_{r-1}} X_{r-1} \xrightarrow{\pi_{r-2}} \cdots \xrightarrow{\pi_1} X_1 \xrightarrow{\pi_0} X_0 = X$$

is a blow-up sequence of order $\geq m$ starting with (X, I, m, E) iff it is a blow-up sequence of order $\geq s$ starting with $(X, W_s(I), s, E)$.

Proof. Everything follows from (3.99) and (3.100), except for the role played by E.

Adding E to (X, I) (resp., to $(X, W_s(I))$) means that now we can use only blow-ups whose centers are in simple normal crossing with E and its total transforms. This poses the same restriction on smooth blow-up sequences for (X, I, E) as on smooth blow-up sequences for $(X, W_s(I), E)$. $\square$

3.12. Order reduction for ideals

In this section we prove the first main implication (3.70.1) of the inductive proof. We start with a much weaker result. Instead of getting rid of all points of order m, we prove only that the set of points of order m moves away from the birational transform of a given divisor E^j.

LEMMA 3.102. *Assume that (3.69) holds in dimensions $< n$. Then for every m, j there is a smooth blow-up sequence functor $\mathcal{BD}_{n,m,j}$ of order m that is defined on triples (X, I, E) with $\dim X = n$, $\max\text{-}\mathrm{ord}\, I \leq m$ and $E = \sum_i E^i$ such that if $\mathcal{BD}_{n,m,j}(X, I, E) =$*

$$\Pi : (X_r, I_r, E_r) \xrightarrow{\pi_{r-1}} (X_{r-1}, I_{r-1}, E_{r-1}) \xrightarrow{\pi_{r-2}} \cdots$$
$$\xrightarrow{\pi_1} (X_1, I_1, E_1) \xrightarrow{\pi_0} (X_0, I_0, E_0) = (X, I, E),$$

then

    (1) $\mathrm{cosupp}(I_r, m) \cap \Pi^{-1}_* E^j = \emptyset$, *and*

    (2) *$\mathcal{BD}_{n,m,j}$ commutes with smooth morphisms (3.34.1) and also with change of fields (3.34.2).*

Assume in addition that there is an ideal $J \subset \mathcal{O}_{E^j}$ such that J is nonzero on every irreducible component of E^j and $\tau_*\left(\mathcal{O}_{E^j}/J\right) = \mathcal{O}_X/I$, where $\tau : E^j \hookrightarrow X$ is the natural injection. Then

(3) $\mathcal{BD}_{n,1,j}(X, I, E) := \tau_* \mathcal{BMO}_{n-1,1}(E^j, J, 1, (E - E^j)|_{E^j})$.

Proof. By (3.101), $W_{m!}(I)$ is D-balanced and order reduction for (X, I, E) is equivalent to order reduction for $(X, W_{m!}(I), E)$. Thus from now on we assume that I is D-balanced.

Let Z_{-1} be the union of those irreducible components $E^{jk} \subset E^j$ that are contained in cosupp(I, m). Let $\pi_{-1} : X_0 \to X$ be the blow-up of Z_{-1}. The blow-up is an isomorphism, but the order of I along E^{jk} is reduced by m and we get a new ideal sheaf I_0. Since max-ord$_{E^{jk}} I \le m$ to start with, max-ord$_{E^{jk}} I_0 = $ max-ord$_{E^{jk}} I - m \le 0$. Thus cosupp(I_0, m) does not contain any irreducible component of E^j.

Next, set $S := E^j$ with injection $\tau : S \hookrightarrow X$, $E_S := (E - E^j)|_S$ and consider the triple $(S, I_0|_S, E_S)$. By the going-up theorem (3.84), every blow-up sequence of order $\ge m$ starting with $(S, I_0|_S, m, (E - E^j)|_S)$ corresponds to a blow-up sequence of order m starting with $(X_0, I_0, E - E^j)$. Since $S = E^j$, every blow-up center is a smooth subvariety of the birational transform of E^j; thus we in fact get a blow-up sequence of order m starting with (X_0, I_0, E). Set

$$\mathcal{BD}_{n,m,j}(X, I, E) := \tau_* \mathcal{BMO}_{n-1,m}(S, I_0|_S, m, E_S) \xrightarrow{\pi_{-1}} X.$$

That is, we take $\mathcal{BMO}_{n-1,m}(S, I_0|_S, m, (E - E^j)|_S)$, push it forward (3.30.3) and compose the resulting blow-up sequence on the right with our first blow-up π_{-1}. (This is the reason for the subscript -1.) By (3.85) we obtain

$$\Pi_r : X_r \to X \quad \text{with} \quad I_r := (\Pi_r)_*^{-1} I, \ E_r := (\Pi_r)_{\text{tot}}^{-1}(E)$$

such that $(\Pi_r)_*^{-1}(E^j)$ is disjoint from cosupp(I_r, m).

The functoriality properties of $\mathcal{BD}_{n,m,j}(X, I, E)$ follow from the corresponding functoriality properties of $\mathcal{BMO}_{n-1,m}(S, I_0|_S, E_S)$. All the steps are obvious, but for the first time, let us go through the details.

Let $h : Y \to X$ be a smooth surjection. Set $E_Y^j := h^{-1}(E^j)$. Then $h|_{E_Y^j} : E_Y^j \to E^j$ is also a smooth surjection and we get the same result whether we first pull back by h and then restrict to E_Y^j or we first restrict to E^j and then pull back by $h|_{E_Y^j}$. That is,

$$\left(h|_{E_Y^j}\right)^* \left(E^j, I_0|_{E^j}, m, (E - E^j)|_{E^j}\right) = \left(E_Y^j, (h^* I)_0|_{E_Y^j}, h^{-1}(E - E^j)|_{E_Y^j}\right).$$

Therefore,

$$\mathcal{BMO}_{n-1,m}\left(E_Y^j, (h^* I)_0|_{E_Y^j}, h^{-1}(E - E^j)|_{E_Y^j}\right)$$
$$= \left(h|_{E_Y^j}\right)^* \mathcal{BMO}_{n-1,m}\left(E^j, I_0|_{E^j}, m, (E - E^j)|_{E^j}\right),$$

and hence

$$h^* \mathcal{BD}_{n,m,j}(X, I, E) = \mathcal{BD}_{n,m,j}(Y, h^* I, h^{-1}(E)).$$

If $h : Y \to X$ is any smooth morphism, we see similarly that the same blow-ups end up with empty centers.

The functoriality property (3.34.2) holds since change of the base field commutes with restrictions.

Assume finally that $\mathcal{O}_X/I = \tau_*(\mathcal{O}_{E^j}/J)$. Every local equation of E^j is an order 1 element in I. Thus $m = \text{max-ord}\, I = 1$ and so $W_{m!}(I) = I$. If J is nonzero on every irreducible component of E^j then $Z_{-1} = \emptyset$ and so $I_0 = I$, proving (3). $\square$

The main theorem of this section is the following.

THEOREM 3.103. *Assume that (3.69) holds in dimensions $< n$. Then for every m there is a smooth blow-up sequence functor $\mathcal{BO}_{n,m}$ of order m that is defined on triples (X, I, E) with $\dim X = n$ and $\text{max-ord}\, I \leq m$ such that if $\mathcal{BO}_{n,m}(X, I, E) =$*

$$\Pi : (X_r, I_r, E_r) \xrightarrow{\pi_{r-1}} (X_{r-1}, I_{r-1}, E_{r-1}) \xrightarrow{\pi_{r-2}} \cdots$$
$$\xrightarrow{\pi_1} (X_1, I_1, E_1) \xrightarrow{\pi_0} (X_0, I_0, E_0) = (X, I, E),$$

then

(1) $\text{max-ord}\, I_r < m$, *and*
(2) $\mathcal{BO}_{n,m}$ *commutes with smooth morphisms (3.34.1) and also with change of fields (3.34.2).*

Assume in addition that there is a smooth hypersurface $\tau : Y \hookrightarrow X$ and an ideal sheaf $J \subset \mathcal{O}_Y$ such that J is nonzero on every irreducible component of Y and $\tau_*(\mathcal{O}_Y/J) = \mathcal{O}_X/I$. Then $\text{max-ord}\, I = 1$ and

(3) $\mathcal{BO}_{n,1}(X, I, \emptyset) = \tau_* \mathcal{BMO}_{n-1,1}(Y, J, 1, \emptyset)$.

The proof is done in three steps.

Step 1 (Tuning I). By (3.101), there is an ideal $W(I) = W_s(I)$ for suitable s, which is D-balanced, MC-invariant and order reduction for (X, I, E) is equivalent to order reduction for $(X, W(I), E)$. (Let us take $s = m!$ to avoid further choices.) Thus from now on we assume that I is D-balanced and MC-invariant.

Step 2 (Maximal contact case). Here we assume that there is a smooth hypersurface of maximal contact $H \subset X$. This is always satisfied in a suitable open neighborhood of any point by (3.80.2), but it may hold globally as well. This condition is also preserved under disjoint unions.

Under a smooth blow-up of order m, the birational transform of a smooth hypersurface of maximal contact is again a smooth hypersurface of maximal contact; thus we stay in the maximal contact case.

We intend to restrict everything to H, but we run into the problem that $E|_H$ need not be a simple normal crossing divisor. We take care of this problem first.

Step 2.1. If $E = \sum_{i=1}^{s} E^i$, we apply (3.102) to each E^i. At the end we get a blow-up sequence $\Pi : X_r \to X$ such that $\text{cosupp}(I_r, m)$ is disjoint from $\Pi_*^{-1} E$.

Note that the new exceptional divisors obtained in the process (and added to E) have simple normal crossings with the birational transforms of H, so $H_r + E_r$ is a simple normal crossing divisor. (I have used the ordering of the index set of E. This is avoided traditionally by restricting (X, I, E) successively to the multiplicity $n - j$ locus of E, starting with the case $j = 0$. The use of the ordering cannot be avoided in (3.111.3), so there is not much reason to go around it here.)

Step 2.2. Once $H + E$ is a simple normal crossing divisor, we restrict everything to the birational transform of H, and we obtain order reduction using dimension induction and (3.102).

Step 3 (Global case). There may not be a global smooth hypersurface of maximal contact $H \subset X$, but we can cover X with open subsets $X^{(j)} \subset X$ such that on each $X^{(j)}$ there is a smooth hypersurface of maximal contact $H^{(j)} \subset X^{(j)}$. Thus the disjoint union

$$H^* := \coprod_j H^{(j)} \subset \coprod_j X^{(j)} =: X^*$$

is a smooth hypersurface of maximal contact. Let $g : X^* \to X$ be the coproduct of the injections $X^{(j)} \hookrightarrow X$.

By the previous step $\mathcal{BO}_{n,m}$ is defined on $(X^*, g^* I, g^{-1} E)$. Then we argue as in (3.37) to prove that $\mathcal{BO}_{n,m}(X^*, g^* I, g^{-1} E)$ descends to give $\mathcal{BO}_{n,m}(X, I, E)$.

Only Steps 2 and 3 need amplification.

3.104 (Step 2, Maximal contact case). We start with a triple (X, I, E), where I is D-balanced and MC-invariant, and assume that there is a smooth hypersurface of maximal contact $H \subset X$. Set $m = \text{max-ord}\, I$.

Warning. As we blow up, we get birational transforms of I which may be neither D-balanced nor MC-invariant. We do not attempt to "fix" this problem, since the relevant consequences of these properties (3.84) and (3.92) are established for any sequence of blow-ups of order m. This also means that we should not pick new hypersurfaces of maximal contact after a blow-up but rather stick with the birational transforms of the old ones.

Step 2.1 (Making $\text{cosupp}(I_r, m)$ and $\Pi_*^{-1} E$ disjoint). To fix notation, write $E = \sum_{i=1}^{s} E^i$, and set $(X_0, I_0, E_0) := (X, I, E)$ and $H_0 := H$. The triple (X_0, I_0, E_0) satisfies the assumptions of Step 2.1.1.

Step 2.1.j. Assume that we have already constructed a smooth blow-up sequence of order m starting with (X_0, I_0, E_0) whose end result is

$$\Pi_{r(j-1)} : X_{r(j-1)} \to \cdots \to X_0, \quad \text{where}$$
$$I_{r(j-1)} := \left(\Pi_{r(j-1)}\right)_*^{-1} I \quad \text{and} \quad E_{r(j-1)} := \left(\Pi_{r(j-1)}\right)_{\text{tot}}^{-1}(E),$$

such that

$$\left(\Pi_{r(j-1)}\right)_*^{-1}(E^i) \cap \text{cosupp}(I_{r(j-1)}, m) = \emptyset \quad \text{for } i < j.$$

Apply (3.102) to $\left(X_{r(j-1)}, I_{r(j-1)}, E_{r(j-1)}\right)$ and the divisor $E^j_{r(j-1)}$ to obtain

$$\Pi_{r(j)} : X_{r(j)} \to X_{r(j-1)} \cdots \to X_0$$

such that

$$\left(\Pi_{r(j)}\right)_*^{-1}(E^i) \cap \text{cosupp}(I_{r(j)}, m) = \emptyset \quad \text{for } i \le j.$$

Note that the center of every blow-up is contained in every hypersurface of maximal contact. Thus $H_{r(j)} := \left(\Pi_{r(j)}\right)_*^{-1} H$ is a smooth hypersurface of maximal contact, and every new divisor in $\left(\Pi_{r(j)}\right)_{\text{tot}}^{-1} E$ is transversal to $H_{r(j)}$. If $E = \sum_{i=1}^s E^i$, then after Step 2.1.s, we have achieved that

- $\left(\Pi_{r(s)}\right)_*^{-1} E$ is disjoint from $\text{cosupp}(I_{r(s)}, m)$, and
- for any hypersurface of maximal contact $H \subset X$, the divisor $H_{r(s)} + E_{r(s)}$ has simple normal crossing along $\text{cosupp}(I_{r(s)}, m)$.

Note that we perform all these steps even if $H + E$ is a simple normal crossing divisor to start with, though in this case they do not seem to be necessary. We would, however, run into problems with the compatibility of the numbering in the blow-up sequences otherwise.

Step 2.2 (Restricting to H). After dropping the subscript $r(s)$ we have a triple (X, I, E) and a smooth hypersurface of maximal contact $H \subset X$ such that $H + E$ is also a simple normal crossing divisor. We can again replace I by $W(I)$ and thus assume that I is MC-invariant. Note that we do not pick a new hypersurface of maximal contact, but use only the birational transforms $H_{r(s)}$ of the old hypersurfaces of maximal contact.

Declare $E^0 := H$ to be the first divisor in $H + E$ and apply (3.102) to $(X, I, H + E)$ with $j = 0$. This gives a sequence of blow-ups $\Pi : X_r \to X$ such that $\text{cosupp} \, \Pi_*^{-1}(I, m)$ is disjoint from $\Pi_*^{-1} H$. However, H is a smooth hypersurface of maximal contact, and hence, by definition, $\text{cosupp} \, \Pi_*^{-1}(I, m) \subset \Pi_*^{-1} H$. Thus $\text{cosupp} \, \Pi_*^{-1}(I, m) = \emptyset$, as we wanted.

Step 2.3 (Functoriality). Assuming functoriality in dimension $< n$, we have functoriality in Step 2.1 by the corresponding functoriality in (3.102).

In Step 2.2 we rely on the choice of a hypersurface of maximal contact H, which is not unique. Let H, H' be two hypersurfaces of maximal contact such that $H + E$ and $H' + E$ are both simple normal crossing divisors. We can use either of the two blow-up sequences $\mathcal{BD}_{n,m,0}(X, I, H + E)$ and $\mathcal{BD}_{n,m,0}(X, I, H' + E)$ to construct $\mathcal{BO}_{n,m}(X, I, E)$.

Here we need that I is MC-invariant. By (3.92) this implies that $(X, I, H + E)$ and $(X, I, H' + E)$ are étale equivalent. Thus the two blow-up sequences $\mathcal{BD}_{n,m,0}(X, I, H + E)$ and $\mathcal{BD}_{n,m,0}(X, I, H' + E)$ are also étale equivalent. By (3.97) this implies that these blow-up sequences are identical.

As we noted in (3.34), the functoriality package is local, so we do not have to consider it separately in the next step.

Step 2.4 (Closed embeddings) Let $\tau : Y \hookrightarrow X$ be a smooth hypersurface and $J \subset \mathcal{O}_Y$ an ideal sheaf such that J is nonzero on every irreducible component of Y and $\tau_*(\mathcal{O}_Y/J) = \mathcal{O}_X/I$. Then I contains the local equations of Y, and so it has order 1. In particular, $I = W(I)$. If $E = \emptyset$ then Step 2.1 does nothing, and in Step 2.2 we can choose $H = Y$. Thus (3.103.3) follows from (3.102.3).

As in (3.37), going from the local to the global case is essentially automatic. For ease of reference, let us axiomatize the process.

THEOREM 3.105 (Globalization of blow-up sequences). *Assume that we have the following:*

(1) *a class of smooth morphisms $\mathcal{M}$ that is closed under fiber products and coproducts (for instance, $\mathcal{M}$ could be all smooth morphisms, all étale morphisms or all open immersions);*

(2) *two classes of triples $\mathcal{GT}$ (global triples) and $\mathcal{LT}$ (local triples) such that*
   (i) *for every $(X, I, E) \in \mathcal{GT}$ and every $x \in X$ there is an $\mathcal{M}$-morphism $g_x : (x' \in U_x) \to (x \in X)$ such that $(U_x, g^*I, g^{-1}E)$ is in $\mathcal{LT}$, and*
   (ii) *$\mathcal{LT}$ is closed under disjoint unions;*

(3) *a blow-up sequence functor $\mathcal{B}$ defined on $\mathcal{LT}$ that commutes with surjections in $\mathcal{M}$.*

Then $\mathcal{B}$ has a unique extension to a blow-up sequence functor $\overline{\mathcal{B}}$, which is defined on $\mathcal{GT}$ and which commutes with surjections in $\mathcal{M}$.

Proof. For any $(X, I, E) \in \mathcal{GT}$ choose $\mathcal{M}$-morphisms $g_{x_i} : U_{x_i} \to X$ such that the images cover X.

Let $X' := \coprod_i U_{x_i}$ be the disjoint union and $g : X' \to X$ the induced $\mathcal{M}$-morphism. By assumption $(X', g^*I, g^{-1}E) \in \mathcal{LT}$.

Set $X'' := X' \times_X X'$. By assumption the two coordinate projections $\tau_1, \tau_2 : X'' \to X'$ are in $\mathcal{M}$ and are surjective.

The blow-up sequence $\mathcal{B}$ for X' starts with blowing up $Z'_0 \subset X'$, and the blow-up sequence $\mathcal{B}$ for X'' starts with blowing up $Z''_0 \subset X''$. Since $\mathcal{B}$ commutes with the τ_i, we conclude that

$$\tau_1^*(Z'_0) = Z''_0 = \tau_2^*(Z'_0). \qquad (3.105.4)$$

If $\mathcal{M} = \{\text{open immersions}\}$, we have proved in (3.37) that the subschemes $Z'_0 \cap U_{x_i} \subset X$ glue together to a subscheme $Z_0 \subset X$. This is the only case we need for the proof of (3.103).

The conclusion still holds for any $\mathcal{M}$, but we have to use the theory of faithfully flat descent; see [Gro95] or [Mur67, Ch.VII].

This way we obtain $X_1 := B_{Z_0}X$ such that $X_1' = X' \times_X X_1$. We can repeat the above argument to obtain the center $Z_1 \subset X_1$ and eventually get the whole blow-up sequence for (X, I, E). $\qquad\square$

The following example, communicated to me by Bierstone and Milman, shows that while principalization proceeds by smooth blow-ups, the resolution of singularities also involves blowing up singular centers.

EXAMPLE 3.106. Consider the subvariety $X \subset \mathbb{A}^4$ defined by the ideal $I = (x^3 - y^2, x^4 + xz^2 - w^3)$. Let us see how the principalization proceeds.

Note that $\operatorname{ord} I = 2$ and $H = (y = 0)$ is a hypersurface of maximal contact. $I|_H = (x^3, xz^2 - w^3)$ has order 3 and $MC(I|_H) = (x, z, w)$. Thus the first step is to blow up the origin in $\mathbb{A}^4$.

Consider the chart $x_1 = x, y_1 = y/x, z_1 = z/x, w_1 = w/x$. The birational transform of I is $I_1 = (x_1 - y_1^2, x_1(x_1 + z_1^2 - w_1^3))$ and $E_1 = (x_1 = 0)$. The order has dropped to 1, so we continue with $(I_1, 1, E_1)$.

Since $\operatorname{cosupp}(I_1, 1)$ is not disjoint from E_1, we proceed as in (3.104.1). The restriction is $I_1|_{E_1} = (x_1, y_1^2)$, and thus next we have to blow up $(x_1 = y_1 = 0)$.

On the other hand, the birational transform of X is

$$X_1 = (x_1 - y_1^2 = x_1 + z_1^2 - w_1^3 = 0).$$

Its intersection with $(x_1 = y_1 = 0)$ is the cuspidal curve $(x_1 = y_1 = z_1^2 - w_1^3 = 0)$. Thus the resolution of X first blows up the origin and then the new exceptional curve, which is singular.

3.13. Order reduction for marked ideals

In this section we prove the second main implication (3.70.2) of the inductive proof. That is, we prove the following.

THEOREM 3.107. *Assume that (3.68) holds in dimensions $\leq n$. Then for every m, there is a smooth blow-up sequence functor $\mathcal{BMO}_{n,m}$ defined on triples (X, I, m, E) with $\dim X = n$ such that if $\mathcal{BMO}_{n,m}(X, I, m, E) =$*

$$\Pi : (X_r, I_r, m, E_r) \xrightarrow{\pi_{r-1}} (X_{r-1}, I_{r-1}, m, E_{r-1}) \xrightarrow{\pi_{r-2}} \cdots$$
$$\xrightarrow{\pi_1} (X_1, I_1, m, E_1) \xrightarrow{\pi_0} (X_0, I_0, m, E_0) = (X, I, m, E),$$

then

(1) $\operatorname{max-ord} I_r < m$,
(2) $\mathcal{BMO}_{n,m}$ *commutes with smooth morphisms (3.34.1) and with change of fields (3.34.2), and*
(3) *if $m = \operatorname{max-ord} I$ then $\mathcal{BMO}_{n,m}(X, I, m, \emptyset) = \mathcal{BO}_{n,m}(X, I, \emptyset)$.*

Before proving (3.107), we show that it implies the two claims in (3.71).

3.108 (Proof of (3.71.1–2)). First, (3.71.1) is the same as (3.107.3).

The claimed identity in (3.71.2) is a local question on X, thus we may assume that there is a chain of smooth subvarieties $Y = Y_0 \subset Y_1 \subset \cdots \subset Y_c = X$ such that each is a hypersurface in the next one. Thus it is enough to prove the case when Y is a hypersurface in X.

Every local equation of Y is in I, thus max-ord $I = 1$. Therefore, $\mathcal{BMO}_{\dim X,1}(X, I, 1, \emptyset) = \mathcal{BO}_{\dim X,1}(X, I, \emptyset)$ by (3.107.3) and (3.103.3) gives that $\mathcal{BO}_{\dim X,1}(X, I, \emptyset) = \tau_* \mathcal{BMO}_{\dim Y,1}(Y, J, 1, \emptyset)$. Putting the two together gives (3.71.2). ☐

3.109 (Plan of the proof of (3.107)). *Step 1.* We start with the unmarked triple (X, I, E), and using (3.68) in dimension n, we reduce its order below m. That is, we get a composite of smooth blow-ups $\Pi_1 : X^1 \to X$ such that $(\Pi_1)_*^{-1} I$ has order $< m$. The problem is that $(\Pi_1)_*^{-1} I$ differs from $(\Pi_1)_*^{-1}(I, m)$ along the exceptional divisors of Π_1, and the latter can have very high order. We decide not to worry about it for now.

Step 2. Continuing with $(X^1, (\Pi_1)_*^{-1}(I, m), (\Pi_1)_{\mathrm{tot}}^{-1} E)$, we blow up subvarieties where the birational transform of (I, m) has order $\geq m$, and the birational transform of I has order ≥ 1.

Eventually we get $\Pi_2 : X^2 \to X$ such that $\mathrm{cosupp}(\Pi_2)_*^{-1} I$ is disjoint from the locus where $(\Pi_2)_*^{-1}(I, m)$ has order $\geq m$. We can now completely ignore $(\Pi_2)_*^{-1} I$. Since $(\Pi_2)_*^{-1} I$ and $(\Pi_2)_*^{-1}(I, m)$ agree up to tensoring with the ideal sheaf of a divisor whose support is in E_r, we can assume from now on that $(\Pi_2)_*^{-1}(I, m)$ is the ideal sheaf of a divisor with simple normal crossing.

Step 3. Order reduction for the marked ideal sheaf of a divisor with simple normal crossing is rather easy.

Instead of strictly following this plan, we divide the ideal into a "simple normal crossing part" and the "rest" using all of E, instead of exceptional divisors only. This is solely a notational convenience.

DEFINITION–LEMMA 3.110. *Given (X, I, E), we can write I uniquely as $I = M(I) \cdot N(I)$, where $M(I) = \mathcal{O}_X(-\sum c_i E^i)$ for some c_i and $\mathrm{cosupp}\, N(I)$ does not contain any of the E^i. $M(I)$ is called the monomial part of I and $N(I)$ the nonmonomial part of I.*

Since the E^i are not assumed irreducible, $\mathrm{cosupp}\, N(I)$ may contain irreducible components of some of the E^i.

3.111 (Proof of (3.107)). We write $I = M(I) \cdot N(I)$ and try to deal with the two parts separately.

Step 1 (Reduction to ord $N(I) < m$). If ord $N(I) \geq m$, we can apply order reduction (3.68) to $N(I)$, until its order drops below m. This happens at some $\Pi_1 : X^1 \to X$. Note that the two birational transforms

$$(\Pi_1)_*^{-1} N(I) \quad \text{and} \quad (\Pi_1)_*^{-1}(I, m)$$

differ only by tensoring with an ideal sheaf of exceptional divisors of Π_1, thus only in their monomial part. Therefore,

$$N\big((\Pi_1)_*^{-1}(I,m)\big) = (\Pi_1)_*^{-1} N(I),$$

and so we have reduced to the case where the maximal order of the non-monomial part is $< m$.

To simplify notation, instead of $(X^1, (\Pi_1)_*^{-1}(I,m), (\Pi_1)_{\text{tot}}^{-1}(E))$, write (X, I, m, E). From now on we may assume that max-ord $N(I) < m$.

Step 2 (Reduction to $\operatorname{cosupp}(I,m) \cap \operatorname{cosupp} N(I) = \emptyset$). Our aim is to continue with order reduction further and get rid of $N(I)$ completely. The problem is that we are allowed to blow up only subvarieties along which (I,m) has order at least m. Thus we can blow up $Z \subset X$ with $\operatorname{ord}_Z N(I) < m$ only if $\operatorname{ord}_Z I \geq m$. We will be able to guarantee this interplay by a simple trick.

Let s be the maximum order of $N(I)$ along $\operatorname{cosupp}(I,m)$. We reduce this order step-by-step, eventually ending up with $s = 0$, which is the same as $\operatorname{cosupp}(I,m) \cap \operatorname{cosupp} N(I) = \emptyset$.

It would not have been difficult to develop order reduction theory for several marked ideals and to apply it to the pair of marked ideals $(N(I), s)$ and (I,m), but the following simple observation reduces the general case to a single ideal:

$$\operatorname{ord}_Z J_1 \geq s \text{ and } \operatorname{ord}_Z J_2 \geq m \Leftrightarrow \operatorname{ord}_Z(J_1^m + J_2^s) \geq ms.$$

Thus we apply order reduction to the ideal $N(I)^m + I^s$, which has order $\geq ms$. Every smooth blow-up sequence of order ms starting with $N(I)^m + I^s$ is also a smooth blow-up sequence of order s starting with $N(I)$ and a smooth blow-up sequence of order m starting with I. Thus we stop after $r = r(m,s)$ steps when we have achieved $\operatorname{cosupp}(I_r, m) \cap \operatorname{cosupp}(N(I_r), s) = \emptyset$. We can continue with $s - 1$ and so on.

Eventually we achieve a situation where (after dropping the subscript) the cosupports of $N(I)$ and of (I,m) are disjoint. Since the center of any further blow-up is contained in $\operatorname{cosupp}(I,m)$, we can replace X by $X \setminus \operatorname{cosupp} N(I)$ and thus assume that $I = M(I)$. The final step is now to deal with monomial ideals.

Step 3 (Order reduction for $M(I)$). Let X be a smooth variety, $\cup_{j \in J} E^j$ a simple normal crossing divisor with ordered index set J and a_j natural numbers giving the monomial ideal $I := \mathcal{O}_X(-\sum a_j E^j)$.

The usual method would be to look for the highest multiplicity locus and blow it up. This, however, does not work, not even for surfaces; see (3.112).

The only thing that saves us at this point is that the divisors E^i come with an ordered index set. This allows us to specify in which order to blow up. There are many possible choices. As far as I can tell, there is no natural or best variant.

Step 3.1. Find the smallest j such that $a_j \geq m$ is maximal. If there is no such j, go to the next step. Otherwise, blow up E^j. Repeating this, we eventually get to the point where $a_j < m$ for every j.

Step 3.2. Find the lexicographically smallest $(j_1 < j_2)$ such that $E^{j_1} \cap E^{j_2} \neq \emptyset$ and $a_{j_1} + a_{j_2} \geq m$ is maximal. If there is no such $(j_1 < j_2)$, go to the next step. Otherwise, blow up $E^{j_1} \cap E^{j_2}$. We get a new divisor, and put it last as E^{j_ℓ}. Its coefficient is $a_{j_\ell} = a_{j_1} + a_{j_2} - m < m$. The new pairwise intersections are $E^i \cap E^{j_\ell}$ for certain values of i. Note that

$$a_i + a_{j_\ell} = a_i + a_{j_1} + a_{j_2} - m < a_{j_1} + a_{j_2},$$

since $a_i < m$ for every i by Step 3.1.

At each repetition, the pair $(m_2(E), n_2(E))$ decreases lexicographically where

$$
\begin{aligned}
m_2(E) &:= \max\{a_{j_1} + a_{j_2} : E^{j_1} \cap E^{j_2} \neq \emptyset\}, \\
n_2(E) &:= \text{number of } (j_1 < j_2) \text{ achieving the maximum.}
\end{aligned}
$$

Eventually we reach the stage where $a_{j_1} + a_{j_2} < m$ whenever $E^{j_1} \cap E^{j_2} \neq \emptyset$.

Step 3.r. Assume that for every $s < r$ we already have the property

$$a_{j_1} + \cdots + a_{j_s} < m \quad \text{if} \quad j_1 < \cdots < j_s \text{ and } E^{j_1} \cap \cdots \cap E^{j_s} \neq \emptyset. \quad (*_s)$$

Find the lexicographically smallest $(j_1 < \cdots < j_r)$ such that $E^{j_1} \cap \cdots \cap E^{j_r} \neq \emptyset$ and $a_{j_1} + \cdots + a_{j_r} \geq m$ is maximal. If there is no such $(j_1 < \cdots < j_r)$, go to the next step. Otherwise, blow up $E^{j_1} \cap \cdots \cap E^{j_r}$, and put the new divisor E^{j_ℓ} last with coefficient $a_{j_1} + \cdots + a_{j_r} - m$. As before, the new r-fold intersections are of the form $E^{i_1} \cap \cdots \cap E^{i_{r-1}} \cap E^{j_\ell}$, where $E^{i_1} \cap \cdots \cap E^{i_{r-1}} \neq \emptyset$. Moreover,

$$a_{i_1} + \cdots + a_{i_{r-1}} + a_{j_\ell} = \left(a_{i_1} + \cdots + a_{i_{r-1}} - m\right) + a_{j_1} + \cdots + a_{j_r},$$

which is less than $a_{j_1} + \cdots + a_{j_r}$ since $a_{i_1} + \cdots + a_{i_{r-1}} < m$ by Step 3.r − 1. Thus the pair $(m_r(E), n_r(E))$ decreases lexicographically, where

$$
\begin{aligned}
m_r(E) &:= \max\{a_{j_1} + \cdots + a_{j_r} : E^{j_1} \cap \cdots \cap E^{j_r} \neq \emptyset\}, \\
n_r(E) &:= \text{number of } (j_1 < \cdots < j_r) \text{ achieving the maximum.}
\end{aligned}
$$

Eventually we reach the stage where the property $(*_r)$ also holds. We can now move to the next step.

At the end of Step 3.n we are done, where $n = \dim X$.

The functoriality conditions are just as obvious as before.

The process greatly simplifies if $m = \text{max-ord}\, I$ and $E = \emptyset$. First, if $E = \emptyset$ then $N(I) = I$. Thus in Step 1 we apply $\mathcal{BO}_{n,m}(X, I, \emptyset)$. Each blow-up has order m, and thus the birational transforms of I agree with the birational transforms of (I, m). At the end of Step 1, $(\Pi_1)_*^{-1} I = (\Pi_1)_*^{-1}(I, m)$.

Thus

$$\operatorname{cosupp}\big((\Pi_1)_*^{-1}(I,m)\big) = \emptyset \quad \text{and} \quad M\big((\Pi_1)_*^{-1}(I,m)\big) = \mathcal{O}_{X^1}.$$

Steps 2 and 3 do nothing, and so $\mathcal{BMO}_{n,m}(X,I,m,\emptyset) = \mathcal{BO}_{n,m}(X,I,\emptyset)$.

$\square$

EXAMPLE 3.112. Let S be a smooth surface, E^1, E^2 two 2 curves intersecting at a point $p = E^1 \cap E^2$ and $a_1 = a_2 = m+1$. Let $\pi : S_3 \to S$ be the blow-up of p with exceptional curve E^3. Then

$$\pi_*^{-1}\big(\mathcal{O}_S(-(m+1)(E^1+E^2)),m\big) = \big(\mathcal{O}_{S_3}(-(m+1)(E^1+E^2)-(m+2)E^3),m\big).$$

Next we blow up the intersection point $E^2 \cap E^3$ and so on. After $r-2$ steps we get a birational transform

$$\big(\mathcal{O}_{S_r}(-\textstyle\sum_{i=1}^r (m+p_i)E^i),m\big),$$

where p_i is the ith Fibonacci number. Thus we get higher and higher multiplicity ideals.

Bibliography

[Abh66]   Shreeram Shankar Abhyankar, *Resolution of singularities of embedded algebraic surfaces*, Pure and Applied Math., Vol. 24, Academic Press, New York, 1966.

[AdJ97]   D. Abramovich and A. J. de Jong, *Smoothness, semistability, and toroidal geometry*, J. Algebraic Geom. **6** (1997), no. 4, 789–801.

[AHV75]   José M. Aroca, Heisuke Hironaka, and José L. Vicente, *The theory of the maximal contact*, Memorias de Matemática del Instituto "Jorge Juan", vol. 29, Instituto "Jorge Juan" de Matemáticas, Consejo Superior de Investigaciones Científicas, Madrid, 1975.

[AHV77]   _____, *Desingularization theorems*, Memorias de Matemática del Instituto "Jorge Juan", vol. 30, Consejo Superior de Investigaciones Científicas, Madrid, 1977.

[AK00]   D. Abramovich and K. Karu, *Weak semistable reduction in characteristic 0*, Invent. Math. **139** (2000), no. 2, 241–273.

[AKMW02]   Dan Abramovich, Kalle Karu, Kenji Matsuki, and Jarosław Włodarczyk, *Torification and factorization of birational maps*, J. Amer. Math. Soc. **15** (2002), no. 3, 531–572 (electronic).

[Alb24a]   G. Albanese, *Transformazione birazionale di una curva algebrica qualunque in un'altra priva di punti multipli*, Rend. della R. Acc. Nazionale dei Lincei **33** (1924), no. 5, 13–14.

[Alb24b]   _____, *Transformazione birazionale di una superficie algebriche in un'altra priva di punti multipli*, Rend. della Circ. Mat. dei Palermo **48** (1924), 321–332.

[AM69]   M. F. Atiyah and I. G. Macdonald, *Introduction to commutative algebra*, Addison-Wesley Publishing Co., Reading, MA, 1969.

[AMRT75]   A. Ash, D. Mumford, M. Rapoport, and Y. Tai, *Smooth compactification of locally symmetric varieties*, Math. Sci. Press, Brookline, MA, 1975.

[Art69]   M. Artin, *Algebraic approximation of structures over complete local rings*, Inst. Hautes Études Sci. Publ. Math. **36** (1969), 23–58.

[Art86a]   _____, *Lipman's proof of resolution of singularities for surfaces*, Arithmetic geometry (Storrs, Conn., 1984), Springer, New York, 1986, pp. 267–287.

[Art86b]   _____, *Néron models*, Arithmetic geometry, Storrs, 1984, Springer, New York, 1986, pp. 213–230.

[AW97]   Dan Abramovich and Jianhua Wang, *Equivariant resolution of singularities in characteristic 0*, Math. Res. Lett. **4** (1997), no. 2–3, 427–433.

[Ben93]   D. J. Benson, *Polynomial invariants of finite groups*, London Math. Soc. Lecture Note Ser., vol. 190, Cambridge Univ. Press, Cambridge, 1993.

[Ber94]   E. Bertini, *Transformazione di una curva algebrica in un'altra con soli punti doppi*, Math. Ann. **44** (1894), no. 1, 79–82.

[BK81]   Egbert Brieskorn and Horst Knörrer, *Ebene algebraische Kurven*, Birkhäuser Verlag, Basel, 1981.

[Bli23]   G. A. Bliss, *The reduction of singularities of plane curves by birational transformation*, Bull. Amer. Math. Soc. (1923), no. 29, 161–183.

[BLR90]   Siegfried Bosch, Werner Lütkebohmert, and Michel Raynaud, *Néron models*, Ergebnisse der Mathematik und ihrer Grenzgebiete (3), vol. 21, Springer-Verlag, Berlin, 1990.

[BM89]   Edward Bierstone and Pierre D. Milman, *Uniformization of analytic spaces*, J. Amer. Math. Soc. **2** (1989), no. 4, 801–836.

[BM91]   _____, *A simple constructive proof of canonical resolution of singularities*, Effective methods in algebraic geometry (Castiglioncello, 1990), Progr. Math., vol. 94, Birkhäuser, Boston, MA, 1991, pp. 11–30.

[BM97]   _____, *Canonical desingularization in characteristic zero by blowing up the maximum strata of a local invariant*, Invent. Math. **128** (1997), no. 2, 207–302.

[BM03]   _____, *Desingularization algorithms. I. Role of exceptional divisors*, Mosc. Math. J. **3** (2003), no. 3, 751–805, 1197.

[BP96]   Fedor A. Bogomolov and Tony G. Pantev, *Weak Hironaka theorem*, Math. Res. Lett. **3** (1996), no. 3, 299–307.

[Bri68]   Egbert Brieskorn, *Rationale Singularitäten komplexer Flächen*, Invent. Math. **4** (1967/1968), 336–358.

[BS00a]   Gábor Bodnár and Josef Schicho, *Automated resolution of singularities for hypersurfaces*, J. Symbolic Comput. **30** (2000), no. 4, 401–428.

[BS00b]   _____, *A computer program for the resolution of singularities*, Resolution of singularities (Obergurgl, 1997), Progr. Math., vol. 181, Birkhäuser, Basel, 2000, pp. 231–238.

[BV01]   A. Bravo and O. Villamayor, *Strengthening the theorem of embedded desingularization*, Math. Res. Lett. **8** (2001), no. 1-2, 79–89.

[Can87]   Felipe Cano, *Desingularization strategies for three-dimensional vector fields*, Lecture Notes in Math., vol. 1259, Springer-Verlag, Berlin, 1987.

[Can04]   _____, *Reduction of the singularities of codimension one singular foliations in dimension three*, Ann. of Math. (2) **160** (2004), no. 3, 907–1011.

[CGO84]   Vincent Cossart, Jean Giraud, and Ulrich Orbanz, *Resolution of surface singularities*, Lecture Notes in Math., vol. 1101, Springer-Verlag, Berlin, 1984.

[CMSB02]   Koji Cho, Yoichi Miyaoka, and N. I. Shepherd-Barron, *Characterizations of projective space and applications to complex symplectic manifolds*, Higher dimensional birational geometry (Kyoto, 1997), Adv. Stud. Pure Math., vol. 35, Math. Soc. Japan, Tokyo, 2002, pp. 1–88.

[Cut02]   Steven Dale Cutkosky, *Monomialization of morphisms from 3-folds to surfaces*, Lecture Notes in Math., vol. 1786, Springer-Verlag, Berlin, 2002.

[Cut04]   _____, *Resolution of singularities*, Graduate Studies in Math., vol. 63, American Math. Soc., Providence, RI, 2004.

[Del71]   Pierre Deligne, *Théorie de Hodge, II*, Inst. Hautes Études Sci. Publ. Math. **40** (1971), 5–57.

[dJ96]   A. J. de Jong, *Smoothness, semi-stability and alterations*, Inst. Hautes Études Sci. Publ. Math. **83** (1996), 51–93.

[Dur79]   Alan H. Durfee, *Fifteen characterizations of rational double points and simple critical points*, Enseign. Math. (2) **25** (1979), no. 1-2, 131–163.

[EH87]   David Eisenbud and Joe Harris, *On varieties of minimal degree (a centen-nial account)*, Algebraic geometry, Bowdoin, 1985, Proc. Sympos. Pure Math., vol. 46, Amer. Math. Soc., Providence, RI, 1987, pp. 3–13.

[EH02]   Santiago Encinas and Herwig Hauser, *Strong resolution of singularities in characteristic zero*, Comment. Math. Helv. **77** (2002), no. 4, 821–845.

[ENV03]   S. Encinas, A. Nobile, and O. Villamayor, *On algorithmic equi-resolution and stratification of Hilbert schemes*, Proc. London Math. Soc. (3) **86** (2003), no. 3, 607–648.

[EV98]   Santiago Encinas and Orlando Villamayor, *Good points and constructive resolution of singularities*, Acta Math. **181** (1998), no. 1, 109–158.

[EV03]   _____, *A new proof of desingularization over fields of characteristic zero*, Rev. Mat. Iberoamericana **19** (2003), no. 2, 339–353.

[FKP05]   Anne Frühbis-Krüger and Gerhard Pfister, *Auflösung von Singularitäten*, Mitt. Dtsch. Math.-Ver. **13** (2005), no. 2, 98–105.

[Fuj90]   Takao Fujita, *Classification theories of polarized varieties*, London Math. Soc. Lect. Note, vol. 155, Cambridge Univ. Press, Cambridge, 1990.

[Ful84]   William Fulton, *Intersection theory*, Ergebnisse der Mathematik und ihrer Grenzgebiete (3), vol. 2, Springer-Verlag, Berlin, 1984.

[Ful93]   _____, *Introduction to toric varieties*, Annals of Math. Stud., vol. 131, Princeton Univ. Press, Princeton, NJ, 1993.

[Gar88]   Paola Gario, *Histoire de la résolution des singularités des surfaces algébriques (une discussion entre C. Segre et P. del Pezzo)*, Cahiers du sém. d'histoire des math., 9, Univ. Paris VI, Paris, 1988, pp. 123–137.

[Gir74]   Jean Giraud, *Sur la théorie du contact maximal*, Math. Z. **137** (1974), 285–310.

[Gir95]   _____, *Résolution des singularités (d'après Heisuke Hironaka)*, Séminaire Bourbaki, 10, Soc. Math. France, Paris, 1995, pp. Exp. No. 320, 101–113.

[GNAPGP88]   F. Guillén, V. Navarro Aznar, P. Pascual Gainza, and F. Puerta, *Hyperrésolutions cubiques et descente cohomologique*, Lecture Notes in Math., vol. 1335, Springer-Verlag, Berlin, 1988.

[GP03]   Pedro D. González Pérez, *Toric embedded resolutions of quasi-ordinary hypersurface singularities*, Ann. Inst. Fourier (Grenoble) **53** (2003), no. 6, 1819–1881.

[GR71]   H. Grauert and R. Remmert, *Analytische Stellenalgebren*, Springer-Verlag, Berlin, 1971, Unter Mitarbeit von O. Riemenschneider, Die Grundlehren der mathematischen Wissenschaften, 176.

[Gro95]   Alexander Grothendieck, *Technique de descente et théorèmes d'existence en géometrie algébrique, I: Généralités. Descente par morphismes fidèlement plats*, Séminaire Bourbaki, 5, Soc. Math. France, Paris, 1995, pp. Exp. No. 190, 299–327.

[Gur03]   R. V. Gurjar, *On a generalization of Mumford's result and related question*, Advances in algebra and geometry (Hyderabad, 2001), Hindustan Book Agency, New Delhi, 2003, pp. 171–178.

[Har77]   Robin Hartshorne, *Algebraic geometry*, Springer-Verlag, New York, 1977.

[Hau03]   Herwig Hauser, *The Hironaka theorem on resolution of singularities (or: A proof we always wanted to understand)*, Bull. Amer. Math. Soc. (N.S.) **40** (2003), no. 3, 323–403 (electronic).

[Hir53]   Friedrich Hirzebruch, *Über vierdimensionale Riemannsche Flächen mehrdeutiger analytischer Funktionen von zwei komplexen Veränderlichen*, Math. Ann. **126** (1953), 1–22.

[Hir64]   Heisuke Hironaka, *Resolution of singularities of an algebraic variety over a field of characteristic zero, I, II*, Ann. of Math. (2) 79 (1964), 109–203; ibid. (2) **79** (1964), 205–326.

[Hir72]   _____, *Schemes, etc.*, Algebraic geometry, Oslo 1970 (Proc. Fifth Nordic Summer School in Math.), Wolters-Noordhoff, Groningen, 1972, pp. 291–313.

[Hir77]   _____, *Idealistic exponents of singularity*, Algebraic geometry (J. J. Sylvester Sympos., Johns Hopkins Univ., Baltimore, MD., 1976), Johns Hopkins Univ. Press, Baltimore, Md., 1977, pp. 52–125.

[Hir83]   _____, *On Nash blowing-up*, Arithmetic and geometry, Vol. II, Progr. Math., vol. 36, Birkhäuser Boston, Mass., 1983, pp. 103–111.

[HK00]   Yi Hu and Sean Keel, *Mori dream spaces and GIT*, Michigan Math. J. **48** (2000), 331–348.

[HLOQ00]   H. Hauser, J. Lipman, F. Oort, and A. Quirós (eds.), *Resolution of singularities*, Progress in Math., vol. 181, Birkhäuser Verlag, Basel, 2000.

[Jel87]   Zbigniew Jelonek, *The extension of regular and rational embeddings*, Math. Ann. **277** (1987), no. 1, 113–120.

[Jun08]   Heinrich W. E. Jung, *Darstellung der Funktionen eines algebraischen Körpers zweier unabhängigen Veränderlichen x, y in der Umgebung einer Stelle $x = a, y = b$*, J. Reine Angew. Math. **133** (1908), 289–314.

[Kal91]   Shulim Kaliman, *Extensions of isomorphisms between affine algebraic subvarieties of k^n to automorphisms of k^n*, Proc. Amer. Math. Soc. **113** (1991), no. 2, 325–334.

[Kar00]   Kalle Karu, *Semistable reduction in characteristic zero for families of surfaces and threefolds*, Discrete Comput. Geom. **23** (2000), no. 1, 111–120.

[Kar05]   _____, *Local strong factorization of toric birational maps*, J. Algebraic Geom. **14** (2005), no. 1, 165–175.

[Kaw06]   Hiraku Kawanoue, *Toward resolution of singularities over a field of positive characteristic Part I.*, arXiv:math.AG/0607009, 2006.

[KKMSD73]   G. Kempf, F. F. Knudsen, D. Mumford, and B. Saint-Donat, *Toroidal embeddings, I*, Springer-Verlag, Berlin, 1973.

[KM83]   J. Kollár and T. Matsusaka, *Riemann-Roch type inequalities*, Amer. J. Math. **105** (1983), no. 1, 229–252.

[KM98]   J. Kollár and S. Mori, *Birational geometry of algebraic varieties*, Cambridge Tracts in Math., vol. 134, Cambridge University Press, Cambridge, 1998, With the collaboration of C. H. Clemens and A. Corti, Translated from the 1998 Japanese original.

[Kol96]   János Kollár, *Rational curves on algebraic varieties*, Ergebnisse der Mathematik und ihrer Grenzgebiete. 3. Folge., vol. 32, Springer-Verlag, Berlin, 1996.

[Kol97]   _____, *Singularities of pairs*, Algebraic geometry—Santa Cruz 1995, Proc. Sympos. Pure Math., vol. 62, Amer. Math. Soc., Providence, RI, 1997, pp. 221–287.

[Kru30]   Wolfgang Krull, *Ein Satz über primäre Integritätsbereiche*, Math. Ann. **103** (1930), no. 1, 450–465.

[KSB88]   J. Kollár and N. I. Shepherd-Barron, *Threefolds and deformations of surface singularities*, Invent. Math. **91** (1988), no. 2, 299–338.

[KSC04]   János Kollár, Karen E. Smith, and Alessio Corti, *Rational and nearly rational varieties*, Cambridge Studies in Advanced Math., vol. 92, Cambridge Univ. Press, Cambridge, 2004.

[Lau83]   Henry B. Laufer, *Weak simultaneous resolution for deformations of Gorenstein surface singularities*, Singularities, (Arcata, Calif., 1981), Proc. Sympos. Pure Math., vol. 40, Amer. Math. Soc., Providence, R.I., 1983, pp. 1–29.

[Laz84]   Robert Lazarsfeld, *Some applications of the theory of positive vector bundles*, Complete intersections (Acireale, 1983), Lecture Notes in Math., vol. 1092, Springer, Berlin, 1984, pp. 29–61.

[Laz04]   _____, *Positivity in algebraic geometry. I-II*, Ergebnisse der Mathematik und ihrer Grenzgebiete. 3. Folge, vol. 48, Springer-Verlag, Berlin, 2004.

[Lip75]   Joseph Lipman, *Introduction to resolution of singularities*, Algebraic geometry (Proc. Sympos. Pure Math., Vol. 29, Humboldt State Univ., Arcata, Calif., 1974), Amer. Math. Soc., Providence, R.I., 1975, pp. 187–230.

[Lip78]   _____, *Desingularization of two-dimensional schemes*, Ann. Math. (2) **107** (1978), no. 1, 151–207.

[Lip83]   _____, *Quasi-ordinary singularities of surfaces in* $\mathbf{C}^3$, Singularities, (Arcata, Calif., 1981), Proc. Sympos. Pure Math., vol. 40, Amer. Math. Soc., Providence, RI, 1983, pp. 161–172.

[Lip00]   _____, *Equisingularity and simultaneous resolution of singularities*, Resolution of singularities (Obergurgl, 1997), Progr. Math., vol. 181, Birkhäuser, Basel, 2000, pp. 485–505.

[Mat70]   Hideyuki Matsumura, *Commutative algebra*, W. A. Benjamin, Inc., New York, 1970.

[Mat89]   _____, *Commutative ring theory*, Cambridge Studies in Advanced Math., vol. 8, Cambridge Univ. Press, Cambridge, 1989.

[Mur67]   J. P. Murre, *Lectures on an introduction to Grothendieck's theory of the fundamental group*, Tata Institute of Fundamental Research, Bombay, 1967.

[Nag62]   Masayoshi Nagata, *Local rings*, Interscience Tracts in Pure and Applied Math., 13, Interscience Publishers, New York-London, 1962.

[Nar83]   R. Narasimhan, *Monomial equimultiple curves in positive characteristic*, Proc. Amer. Math. Soc. **89** (1983), no. 3, 402–406.

[Nér64]   André Néron, *Modèles minimaux des variétés abéliennes sur les corps locaux et globaux*, Inst. Hautes Études Sci. Publ. Math. **21** (1964), 128.

[New60]   Isaac Newton, *The correspondence of Isaac Newton, Vol. II: 1676–1687*, Cambridge Univ. Press, New York, 1960.

[Noe71]   M. Noether, *Ueber die algebraischen Functionen einer und zweier Variabeln*, Nachr. von der Königl. Ges. der Wiss. und der G. A. Universität Göttingen **9** (1871), 267–278.

[Nöt75]   M. Nöther, *Ueber die singulären Werthsysteme einer algebraischen Function und die singulären Punkte einer algebraischen Curve*, Math. Ann. **9** (1875), no. 2, 166–182.

[Oda88]   Tadao Oda, *Convex bodies and algebraic geometry*, Ergebnisse der Mathematik und ihrer Grenzgebiete (3), vol. 15, Springer-Verlag, Berlin, 1988.

[Par99]   Kapil H. Paranjape, *The Bogomolov-Pantev resolution, an expository account*, New trends in algebraic geometry (Warwick, 1996), London Math. Soc. Lecture Note Ser., vol. 264, Cambridge Univ. Press, Cambridge, 1999, pp. 347–358.

[Rei88]   Miles Reid, *Undergraduate algebraic geometry*, London Math. Soc. Student Texts, vol. 12, Cambridge Univ. Press, Cambridge, 1988.

[Rei00]   Heinrich Reitberger, *The turbulent fifties in resolution of singularities*, Resolution of singularities (Obergurgl, 1997), Progr. Math., vol. 181, Birkhäuser, Basel, 2000, pp. 533–537.

[Rie90]   Bernhard Riemann, *Gesammelte mathematische Werke, wissenschaftlicher Nachlass und Nachträge*, Springer-Verlag, Berlin, 1990.

[Sal79]   Judith D. Sally, *Cohen-Macaulay local rings of maximal embedding dimension*, J. Algebra **56** (1979), no. 1, 168–183.

[Sei68]   A. Seidenberg, *Reduction of singularities of the differential equation $A\,dy = B\,dx$*, Amer. J. Math. **90** (1968), 248–269.

[Ser56]   Jean-Pierre Serre, *Géométrie algébrique et géométrie analytique*, Ann. Inst. Fourier, Grenoble **6** (1955–1956), 1–42.

[Ser79]   ———, *Local fields*, Graduate Texts in Mathematics, vol. 67, Springer-Verlag, New York, 1979.

[Sha94]   Igor R. Shafarevich, *Basic algebraic geometry. 1–2*, Springer-Verlag, Berlin, 1994.

[Smi95]   Larry Smith, *Polynomial invariants of finite groups*, Res. Notes in Math., vol. 6, A K Peters Ltd., Wellesley, MA, 1995.

[ST54]   G. C. Shephard and J. A. Todd, *Finite unitary reflection groups*, Canadian J. Math. **6** (1954), 274–304.

[Tei82]   Bernard Teissier, *Variétés polaires. II. Multiplicités polaires, sections planes, et conditions de Whitney*, Algebraic geometry (La Rábida, 1981), Lecture Notes in Math., vol. 961, Springer, Berlin, 1982, pp. 314–491.

[vdW91]   B. L. van der Waerden, *Algebra. Vols. I-II*, Springer-Verlag, New York, 1991.

[Vil89]   Orlando Villamayor, *Constructiveness of Hironaka's resolution*, Ann. Sci. École Norm. Sup. (4) **22** (1989), no. 1, 1–32.

[Vil92]   ———, *Patching local uniformizations*, Ann. Sci. École Norm. Sup. (4) **25** (1992), no. 6, 629–677.

[Vil96]   ———, *Introduction to the algorithm of resolution*, Algebraic geometry and singularities (La Rábida, 1991), Progr. Math., vol. 134, Birkhäuser, Basel, 1996, pp. 123–154.

[Vil00]   ———, *On equiresolution and a question of Zariski*, Acta Math. **185** (2000), no. 1, 123–159.

[Vil06]   ———, *Rees algebras on smooth schemes: integral closure and higher differential operators*, arXiv:math.AC/0606795, 2006.

[Wal35]   Robert J. Walker, *Reduction of the singularities of an algebraic surface*, Ann. of Math. (2) **36** (1935), no. 2, 336–365.

[Wło00]   Jarosław Włodarczyk, *Birational cobordisms and factorization of birational maps*, J. Algebraic Geom. **9** (2000), no. 3, 425–449.

[Wło03]   ———, *Toroidal varieties and the weak factorization theorem*, Invent. Math. **154** (2003), no. 2, 223–331.

[Wło05]   ———, *Simple Hironaka resolution in characteristic zero*, J. Amer. Math. Soc. **18** (2005), no. 4, 779–822 (electronic).

[Zar39]   Oscar Zariski, *The reduction of the singularities of an algebraic surface*, Ann. of Math. (2) **40** (1939), 639–689.

[Zar44]   ———, *Reduction of the singularities of algebraic three dimensional varieties*, Ann. of Math. (2) **45** (1944), 472–542.

[Zar71]   ———, *Algebraic surfaces*, Springer-Verlag, New York, 1971.

Index